INSPECTING COMMERCIAL, INDUSTRIAL, AND RESIDENTIAL CONSTRUCTION

G. L. TAYLOR

D1768178

McGRAW-HILL

New York   Chicago   San Francisco   Lisbon   London
Madrid   Mexico City   Milan   New Delhi   San Juan
Seoul   Singapore   Sydney   Toronto

1 2 3 4 5 6 7 8 9 0   IMP/IMP   0 1 0 9 8 7 6 5 4

ISBN 0-07-144889-6

The sponsoring editor for this book was Cary Sullivan, the editing supervisor was Suzanne Ingrao, and the production supervisor was Sherri Souffrance. It was set in Futura Book by North Market Street Graphics. The art director for the cover was Anthony Landi.

Printed and bound by Imago.

McGraw-Hill books are available at special quantity discounts to use as premiums and sales promotions, or for use in corporate training programs. For more information, please write to the Director of Special Sales, McGraw-Hill Professional, Two Penn Plaza, New York, NY 10121-2298. Or contact your local bookstore.

This book is printed on acid-free paper.

The IBC Building Code's Purpose

"101.3 Intent. The purpose of this code is to provide minimum requirements to safeguard the public health, safety and general welfare. . . ."

INTERNATIONAL CODE COUNCIL

CONTENTS

CHAPTER 1
INTRODUCTION

Up to Code Inspectors Guides comprise a series of manuals covering different CSI construction procedures and standards for commercial and industrial projects. Although based on the IBC 2003 and IRC 2003 codes, these guides are not intended to replace these codes or any of the other model codes and/or specifications in the contract documents. The goal is for the Inspector and/or project site managers to use these guides as a basis for establishing his or her inspection guidelines, and for providing the client with a well-built project meeting the specifications. The convenient sizing of these guides allows the Inspector to carry them around in a pocket, making code and standards information readily accessible. Furthermore, our intent is for this reference tool to be instrumental in the construction of better buildings and to be a valuable training guide for those just entering the construction field. These guides are valuable tools for architects, engineers, project managers, tradespeople, and contractors, as well as inspectors.

REFERENCED MATERIALS

- International Building Code 2003
- International Residential Code 2003
- National Electrical Code (NEC) 2002
- International Fuel Gas Code 2003
- International Fire Codes 2003
- International Energy Conservation Code 2000
- International Mechanical Code 2003
- ACI 318-02/318R-02 Building Code Requirements for Structural Concrete and Commentary
- ACI 530/530.1-02/530R/530.1R-02 Building Code Requirements and Commentary for Masonry Structures and Specification for Masonry Structures and Related Commentaries
- ACI 301-99 Specifications for Structural Concrete for Buildings
- ACI 306.1-90 Standard Specification for Cold Weather Concrete
- ACI 305R-99 Hot Weather Concreting

- ACI 302.1R-96 Guide for Concrete Floor and Slab Construction
- ACI 117-90/177R-90 Standard Tolerances for Concrete Construction and Materials (AC1117-90) and Commentary (ACI117R-90)
- ACI SP-2-99 Manual of Concrete Inspection
- ACI 311.4R-00 Guide for Concrete Inspection
- American Concrete Institute (ACI) 347R "Formwork for Concrete"
- Americans with Disability Act of 1990 (ADA)
- American Forest Association
- American National Standards Institute (ANSI)
- American Society for Testing and Materials (ASTM)
- American Wood Preservers Association (AWPA)
- APA—The Engineered Wood Association (EWA)

One or more International Codes currently enforced statewide

One or more International Codes enforced within state at local level

Adopted statewide with future enforcement date

Figure 1.1   ICC CODE ADOPTIONS

- Factory Mutual (FM)

- Truss Plate Institute (TPI) HIB-91

- OSHA Safety and Health (29 CFR 1926), 2003

- Portland Cement Association (PCA) Design and Control of Concrete Mixtures

- American Society of Concrete Contractors (ASCC) "Contractor's Guide to Quality Concrete Construction"

THE INSPECTOR'S ROLE

As an Owner's Representative and/or the Project Resident Inspector, you should be thoroughly familiar with all the contract documents, including the plans with all changes, specifications, and contracts submittals such as shop drawings. Plans and specifications should include all revisions, changes, and amendments. In addition, you should be thoroughly familiar with the project's reporting requirements as well as the specific duties and responsibilities (including the limits) associated with the project. Procedures and responsibilities will differ from project to project. It is crucial that the Inspector have a clear understanding of the project's responsibilities, as well as all reporting required prior to the start of the project.

Responsibilities of the Inspector

Inspectors have different responsibilities and authorities, depending on the organizational setup, and size and scope of the project. Each Inspector should be clear on the answers to each of the following questions:

❏ Do I understand the limits of my responsibility?
❏ Do I have the technical knowledge required for this project? Can any gaps in my knowledge be effectively covered with the help of other inspection staff and or consultants?
❏ Do I fully understand all reporting procedures?
❏ Do I fully understand where I fit in the organizational chart and to whom I report?
❏ Am I given sufficient authority to carry out my duties and responsibilities?
❏ Make sure you know the extent of your authority. At a minimum, your responsibility is to inspect all work and ensure that it's accomplished in accordance with the contract plans and specifications. Be sure you have the authority you need to meet that responsibility.
❏ Check with your supervisor with respect to your authority to stop subcontractor operations for safety violations, construction deficiencies, or other potential problems.

Quality Control Issues in the Inspector's Role

• Subcontractors may be required to provide additional inspectors in the interest of quality control either part time or full time, depending on the requirements of the project. This is especially true for government projects. As the project's primary Inspector, it is your responsibility to oversee any additional inspectors, and to determine their qualifications and ability to perform their duties.

- The main role of the Inspector is to ensure the owner that the quality requirements of the contract are satisfied.

- At times, projects will require a Subcontractor Quality Control Program, which is made up of inspectors responsible for the quality of each subcontracted aspect of the job. For example, the masonry subcontractor would have a masonry inspector, the electrician would have an electrical inspector, etc. Although the responsibility for overseeing a Subcontractor Quality Control Program is usually the contractor's, you should develop a close working relationship with each Inspector. For some projects, they may report directly to you.

- Effective project inspections require a serious and concentrated effort on the part of all the Inspectors, as well as all site management personnel.

CONTRACT REQUIREMENTS

Contract requirements provide the tools for the accomplishment of the goals. Before the start of construction, the Project Manager shall conduct a meeting of all contractors and discuss their individual quality control plans and procedures. Construction should not start until the meeting has taken place, and, at minimum, until the Project Quality Control plan has been submitted and accepted. Project staffing should be sufficient to obtain the quality of construction designed in the plans and specifications.

The following sections describe the methods for meeting contract requirements.

Preconstruction Meetings

These meetings may be held before each stage of construction. For example, at the start of foundations, masonry work, slab on grade, plumbing grounds, etc., ensure the following:

- ❑ The requirements are understood by all managers and workers.
- ❑ The documentation is complete.
- ❑ The materials are on hand.
- ❑ The people who are to perform the work understand what will be considered satisfactory workmanship.

Both the contract specifications and technical standards referenced in the contract specifications must be in the Job Site Trailer library and available to the inspectors. For instance, the truss placement specifications depend entirely on the Truss Plate Institute HIB-91 (TPI) Specifications for all requirements. If the Inspector doesn't have the TPI specifications, he cannot know or enforce these provisions.

Initial Inspections

These inspections must be conducted in a timely manner at the beginning of a definable feature of work. A check of the preliminary work will determine whether or not the subcontractor—through his Quality Control Manager and the craftsmen involved—thoroughly understands and is capable of accomplishing the work as specified. Check for proper implementation of safety procedures according to the approved Safety Plan at this time.

Follow-up Inspections

Follow-up inspections are conducted daily when work is in progress. This ensures that the controls established in the earlier phases of inspection continue to conform to the contract requirements.

In all projects there is work that is eventually covered and cannot be inspected after the fact.

This includes concrete, where the size, number, and location of reinforcing steel cannot be readily determined after the concrete is placed. Underground utilities cannot be inspected after covering. Work of this nature must be closely controlled and monitored during construction. If the contractor was notified to not cover until you have completed your inspection but does so anyway, then you can direct him to uncover the work at his expense!

PLANS AND SPECIFICATIONS

Review of Plans and Specifications

Make a thorough review of plans and specifications before the bidding period.

❑ Watch for omissions.
❑ Watch for discrepancies between plans and specifications.
❑ Check plans and specifications against requirements that have been problematic on similar jobs.
❑ Compare elevations, grades, and details shown on plans as existing against those at the actual site.
❑ Report all errors, omissions, discrepancies, and deficiencies to the Project Manager.

Marked and Posted Plans and Specifications

Always keep a posted and marked up set of plans and specifications convenient for ready reference. Keep them up to date!

❏ Make sure that the Subcontractor has this same information.
❏ Anticipate work operations by reviewing the plans and specifications for each operation before it begins.
❏ Discuss contract requirements with the Subcontractor before each construction phase begins.
❏ Highlight and/or make notes of those provisions which need special attention, such as:
    (a) Unusual requirements, such as additional concrete test.
    (b) Those requirements overlooked by other contractors.
    (c) Repetitive deficiencies.
    (d) Conflicting specifications and drawings.
    (e) Code violations.
    (f) Use the checklists in the Design Quality Control chapter to help find significant items in the plans and specifications.

REQUIRED GEOGRAPHIC DESIGN CRITERIA

Ground snow load	
Roof snow load	
Wind speed	
Seismic zone	
Weathering zone	
Frost line	
Termite zone	
Decay zone	
Flood zone	
Winter design temperature	
Climate zones (energy code)	
Heating degree days	
Cooling degree days	
Radon zone	
Exposure profile (wind)	
January average temperature	

Shop Drawings

❏ Review the prepared subcontractor submittal register, plans, and specifications. Check submittal register for inclusion of all shop drawings required, including layouts of equipment, equipment rooms, etc. The Inspector must have copies of all shop drawings!

- ❏ The Subcontractor is required to enter his or her data onto the submittal register and submit it to the Project Control Administrator or others tasked with this responsibility. Compare this submittal with your check list.
- ❏ The Project Manager is required to periodically review and update the submittal register. The Inspector should monitor each change.
- ❏ Check the submittal register to prevent untimely and omitted submittals so as to avoid delay of construction. Check specifications for required turnaround time requirements.
- ❏ Compare the shop drawings to the contract requirements and report apparent differences to your supervisor.
- ❏ Make sure each detail on the shop drawing is clearly presented.
- ❏ The Subcontractor must make notes on his submittal of items that deviate from contract requirements.
- ❏ Check material being installed against the approved shop drawing. (If the Subcontractor installs unapproved material, inform him or her in writing that the material, if not subsequently approved, will be removed and replaced at his or her expense.)

Inspection Report

(a) Prepare a complete and accurate daily inspection report. Modify the form to reflect all requirements noted in the specifications and contract documents. Include the following:

❑ **Conditions** weather, moisture, soil conditions, etc. Note when and how an adverse site condition hampered or shut down a particular operation.

❑ **Activities** work phases, including locations and descriptions of each activity and the inspection.

❑ **Controversial issues** disputes, questionable items, etc. (Also, note if they were settled and, if so, how they were settled.)

❑ **Deficiencies and violations** description, location, and corrective action.

❑ **Instructions** given and received; identify recipient and source.

❑ **Progress information** report all delays, anticipated and actual, and action taken or contemplated.

❑ **Equipment** report arrival and departure of each major item of equipment by manufacturer, model, serial number, and capacity; report equipment in use and idle equipment.

- ❏ **Reports** make sure reports are identified, dated, and signed.
- ❏ **Safety** check the daily report each day for accuracy and to ensure that instructions received are noted.

PRECONSTRUCTION MEETING

Attendees

Both the Inspector and the Project Manager should attend this conference as well as all contractors' representatives associated with the project.

Documentation of Meeting

Minutes of the meeting should be available to each quality assurance/quality control representative. The subject of the proposed quality control plan should be well documented.

EQUIPMENT PROPOSAL/EVALUATION

The following issues must be resolved and/or finalized before start of construction.

- ❏ Does equipment proposed by the Subcontractor have proper approval for use?
- ❏ Certain equipment requires a safety test or check before initial operation at the site.

- ❏ Some equipment requires a permit or license before use.
- ❏ Have daily/weekly equipment rates been approved?
- ❏ Has equipment been recently inspected?
- ❏ Have all oils, lubricants, and their containers been properly discarded per EPA and OSHA requirements?

CLAIMS AND DISPUTES

Be sure the following conditions are met:

- ❏ Always be alert to possible claims or matters of possible dispute.
- ❏ When you discover that a claim or dispute is imminent, notify your supervisor and record all facts in your (Inspector) daily reports.
- ❏ Make sure that adequate and accurate records of facts, materials, labor, and equipment associated with the claim or dispute are on file.
- ❏ Situation photographs should be taken to supplement the record.
- ❏ Differing site conditions may be cause for a claim. Subcontractors must notify PM in writing before disturbing conditions.

PROGRESS SCHEDULES

Steps to ensure efficient evaluation are as follows:

- Assist the Subcontractor as he or she prepares initial and revised progress schedules.
- Be certain the Contractor submits timely updates.
- Be familiar with the approved progress schedule, carefully watching for any slippage in progress.
- Anticipate slowdowns and delays affecting progress.
- Promptly report all delays to the Project Manager and record them in the daily reports. Perform manpower analysis as needed.
- When construction falls behind schedule, carefully examine the construction operations for ways to improve efficiency and report your findings to the Superintendent and Project Manager.
- Be very careful not to presume direction of the Subcontractor's operation (the PM/Superintendent is responsible to direct the Contractor on how to improve his progress).
- Monitor required contract milestones and the final completion date.

LABOR ENFORCEMENT

The following precautions must be taken to avoid labor liabilities:

- Keep informed of the labor requirements of the contracts.
- Avoid taking part in any labor disputes. Inform the Project Manager of any labor disputes.
- Check that required posters and minimum wage rates are kept in a conspicuous place.
- Make spot checks with Sub-subcontractors' employees to verify that Davis-Bacon wage rates are being paid for the work classification being performed (if required).

STORAGE OF MATERIALS

- ❑ Check that adequate space is available for the Subcontractors' operations and storage areas (before the materials are scheduled for delivery).
- ❑ Check that approval has been obtained for temporary sheds, buildings, etc. that the Subcontractor proposes to install.
- ❑ See that material and equipment are properly stored and protected.
- ❑ Check that safety requirements, especially in the storage of flammable or explosive materials, are adhered to.

- ❑ Check that temporary structures are secured against wind damage.
- ❑ Check that the necessary heating and ventilating are provided.

SUB-SUBCONTRACTOR'S PAYMENT ESTIMATES (IF REQUIRED)

- ❑ Check work evaluation and payment specifications for each item of work to be accomplished.
- ❑ Review the schedules of values and methods of measurement for payment.
- ❑ Assist the Project Manager in preparing pay estimates.
    - (a) Record timely measurements of work completed and accomplished each pay period.
    - (b) Keep orderly, neat, and accurate records of measurements.
- ❑ Check material on hand for which payment is being made for:
    - (a) Fair market value of materials.
    - (b) Conformance with contract requirements (see submittal).
    - (c) Proper storage and protection.
    - (d) Reduction in quantity by material was placed in the work.

❑ Monitor all increases or decreases in the quantity of work shown on the unit price schedules.

(a) Make as accurate an estimate as possible of variations in quantities.

(b) Report these variations in quantities promptly to the Construction and/or Project Manager.

(c) Keep all estimates for future record.

RIGHTS-OF-WAY

Check that all rights-of-way are obtained before beginning construction or entering the property.

(a) Require written evidence if Subcontractor-obtained.

(b) Know the limits of rights-of-way and locations of benchmarks that may be used to determine location and elevations.

(c) Post signs for workers and drivers to mark limits of operational area.

PHOTOGRAPHS

Property Evaluation and Overview

Photographs will provide information that can evaluate potential hazardous conditions as well as an overview of work progress. Check them for the following indications:

- ❏ Views of major construction projected during various stages of progress.
- ❏ Materials or construction related to changed conditions, claims, or potential claims.
- ❏ Work in place for which removal has been ordered because of noncompliance with plans and specifications.
- ❏ Photos of technical interest.
- ❏ Bad and good safety practices by the contractors.
- ❏ New methods of construction.
- ❏ Property or material damages.
- ❏ Manufacturers' labels and installation instructions.
- ❏ Emergency conditions and safety violations.
- ❏ Accident scenes.
- ❏ Defective work
    (1) Check that each picture taken is completely described, identified, and dated.
    (2) When possible, use a tape measure or other measurement device in pictures to show actual sizes and distances.

RECORD DRAWINGS

(a) The Record Drawings should be reviewed monthly by the Project Manager to ensure their accuracy.

(b) The Site Superintendent or Project Manager must ensure that as soon as a change or addition is made in construction it is noted on the Record Drawing. In some cases, however, this becomes the Inspector's responsibility. Good inspection practice dictates that the Inspector keep good record drawings whether or not they are the set to be turned over to the client or owner.

(c) The following items must be considered in the changes for Record Drawings:

- Size, type, and location of existing and new utility lines.
- Layout and schematic drawings of electrical circuits and piping; include sleeve drawings and diagrams.
- Dimensions and details transferred from shop drawings.
- Final survey records of cross sections, borrow pits, and layout of all earthwork.
- Actual locations of anchors, construction and control joints, etc. in concrete, where they are different from those shown on Contract drawings.
- Changes in equipment location and architectural features.
- Any and all project Change Orders and Field Directives.

PROJECT TURNOVER

The Inspector may have the responsibility of providing the client or owner the official turnover documents (OEM manuals). The following records and materials will be needed.

- ❏ Record of property name, make, and model number of each piece of equipment.
- ❏ All equipment test reports.
- ❏ Approved shop drawings.
- ❏ Operating and maintenance instructions.
- ❏ Spare parts and tools.
- ❏ Keys.
- ❏ Guarantees with required contract and expiration date.
- ❏ Record (As-Built) Drawings.

Check meeting minutes and contract documents for any additional requirements.

QUALITY CONTROL

Remember that the Inspector's responsibilities begin at the inception of construction and end only with the final acceptance by the owner. The Inspector's primary objective is to verify what has been accomplished as well as possible oversights.

SAFETY

(a) The overall Project Safety Programs as well as each individual contractor's safety program must be approved and enforced every day. This enforcement is usually not the duty of the Inspector, unless specified in the contract. Large projects will have a full-time Safety Manager.

(b) Fully assess all work or operations for safety compliance before proceeding with inspecting for the technical compliance.

(c) Be familiar with each contractor's accident prevention programs. These plans should be discussed and finalized before any construction begins.

(d) Plan to attend a different contractor's weekly safety meeting. Stand ready to evaluate and advise.

(e) Applicable Occupational Safety and Health Act (OSHA).

INSPECTION FILES

The following list suggests those files that the Inspector will require for a project. Depending on the size and complexity of the project, the Inspector may need to add additional files, as appropriate, to ensure adequate documentation for the project.

(a) GENERAL
- ❏ Project Contract
- ❏ Contractors Contract
- ❏ Clarifications
- ❏ Request for Change Orders
- ❏ Approved Change Orders
- ❏ Field Directives
- ❏ Claims
- ❏ Schedule of Values
- ❏ Request for Payments
- ❏ Owner furnished Labor/Materials
- ❏ Selections
- ❏ Shop Drawings

(b) CORRESPONDENCE
- ❏ Architect/Engineer
- ❏ Client
- ❏ Contractor
- ❏ Testing labs
- ❏ Consultants
- ❏ Others/Misc.

(c) GOVERNMENT AGENCIES
- ❏ Permits
- ❏ Fire Marshal
- ❏ Certified Payrolls
- ❏ Special

(d) FIELD

- ❏ Transmittals
- ❏ Sketches
- ❏ Request for Information (RFI)
- ❏ Meeting Minutes
- ❏ Schedules
- ❏ Issues Log
- ❏ Daily/Weekly Reports
- ❏ Safety Plans/Reports

(e) TECHNICAL INFORMATION

- ❏ One file per CSI Division

(f) CLOSE-OUTS

- ❏ MSDA Sheets
- ❏ Equipment Instructions
- ❏ Certificate of Occupancy
- ❏ Code Inspection Reports
- ❏ Warrantees/Guarantees
- ❏ Record Drawings

Recommended Equipment

- ❏ 12-ft steel tape measure
- ❏ 100-ft cloth/plastic tape measure
- ❏ 4-ft level
- ❏ 8-ft level
- ❏ Scales (architect/engineer)
- ❏ Pocket calculator
- ❏ Flashlight
- ❏ Penlight

- ❏ Camera
- ❏ Speed square
- ❏ Magnifying glass
- ❏ Thermometer
- ❏ Thickness gauge
- ❏ Protective clothing
- ❏ Hard hat
- ❏ Safety glasses
- ❏ Spud wrench
- ❏ Circuit tester
- ❏ Voltmeter
- ❏ Ampmeter
- ❏ Wire gauge
- ❏ Depth gauge

THINK SAFETY AT ALL TIMES

CHAPTER 2
DESIGN QUALITY CONTROL

DESIGN QUALITY CONTROL CHECKLISTS

It is estimated that more than 50% of the problems encountered on any given construction project could have been avoided if proper review and quality control procedures were implemented <u>before</u> the issuance of the plans or specifications. Most often, designers do not even read their own specifications. The most successfully implemented projects are those that allow time for thorough reviews and identification of potential problems. Listed next are some frequently encountered problems that arise from contract documents. Many are simply common sense; however, failure to adhere to them can result in project failures that could have been (and should have been) avoided!

OVERVIEW

- Work "by others" and work "this contract" are clearly differentiated and interface points identified.

- All known existing features and improvements are properly and completely delineated and dimensioned.
- Orientation, horizontal coordinate systems, elevations, and vertical datum are properly shown and referenced.
- Adequate subsurface investigations of the site have been made and logs and notes thereof are clearly shown on plans and referred to in specifications.
- The recommendations of the Geotechnical Report have been considered in establishment of control elevations, foundation treatment, and assignment of bearing values for footing design. (Who has ownership and responsibility for complying with the recommendations?)
- Adequate provisions have been made in the specifications for protection and maintenance of, access to, and utility services for existing facilities.
- All documents have been logically ordered and a table of contents provided.
- All documents, specifications, and plans have been dated and <u>stamped by the designer</u>!

- The scale and orientation of the drawings are consistent throughout the complete set of drawings.
- The Statement of Work (SOW) shown in the Request for Quotation (RFQ) has been passed through to the current design directives to the individual subcontractors.
- Annotated, approved comments from previous reviews, as well as correspondence and all meeting minutes, are included in the design.

PLANS AND SPECIFICATIONS

❑ All necessary details, notes, schedules, and dimensions are shown on the drawings and are fully consistent throughout.

Civil Details Required

- Gutter
- Storm drainage
- Drainage schedule
- Erosion control
- Manholes
- Meter/water vault
- Gas lines
- Oil separator
- Fire loop w/PIV, hydrants

- Steam/condensation lines
- Area to be cleared
- Fence and gates (size of post, gate type, and widths)
- Demolition areas
- Typical pavements
- Bollard locations
- Misc concrete pads
- Landscape plan and schedule

❑ Title blocks, drawing titles, drawing scales, and specification subtitles and section identification markings are shown and referenced.

❑ Requirements for installation of owner-furnished equipment are clearly delineated.

❑ Ample space allowances are available for installation and servicing of equipment.

❑ The terminology used on the drawings agrees with that used in the specifications and does not repeat requirements stated in the specifications.

❑ Finish and color schedules have been coordinated with drawings.

❑ When drawings are printed at full size, all lettering, dimensions, symbols, wiring and piping runs, etc. are clear and distinct.

- ❏ The drawings and specifications for all disciplines have been properly reviewed and coordinated to preclude conflicts.
- ❏ Complete legends for each discipline, including all symbols, are shown on the plans.
- ❏ North arrow and graphic/bar scales are shown correctly on all site plans.

CIVIL/SITE DESIGN CHECKLISTS

Civil Design

- ❏ Existing and finished grades are shown.
- ❏ Haul routes, disposal/borrow sites, construction contractor's storage area, construction limits, and construction staging area are shown.
- ❏ Existing utilities, sizes, and materials are shown.
- ❏ New underground utilities have been checked for conflicts against the site plans.
- ❏ Utility tie-in locations agree with mechanical stub-out plan.
- ❏ Profile sheets show underground utilities and avoid conflicts between new and existing.
- ❏ Property lines and limits of clearing, grading, turfing, or mulch have been shown and are consistent with architectural and/or landscaping plans.

- ❑ Fire hydrant and power/telephone pole locations correspond with electrical and architectural drawings.
- ❑ Basis of horizontal and vertical control is given, and the control points are located properly with pertinent data shown (for example, BM's/CP's elevations).
- ❑ Tops of valve boxes and manholes match finished grades, pavement, swales or sidewalks.
- ❑ Boring locations, soil classifications, water table, and depth of rock are shown on the plans or in the write-ups.
- ❑ Rigid pavement joint plans are shown with reasonable spacing.
- ❑ Foundation coordinates are shown on the foundation plan and coordinated with architectural drawings.
- ❑ Finished floor elevations match on architectural and structural drawings.
- ❑ Civil specifications are coordinated with plans.
- ❑ Storm and sewage drains from the facility have adequate capacity.
- ❑ Directions to contractors are not contradicted in plan notes and in the specifications.
- ❑ Removal, demolition plan(s) is (are) complete.

DESIGN

- ❏ Construction limit line is shown, including removal of existing pavements when required.
- ❏ Sufficient attention has been given to preserving the natural terrain and trees.
- ❏ Sufficient general notes, dimensions, and elevations are shown for proper construction layout, including construction baseline (B/L) on finish grade spot elevations are indicated on graded earth areas and along pavements on "Grading and Paving Plan."
- ❏ Slopes of paved surfaces and graded earth areas are satisfactory and within criteria of maximum and minimum grades to prevent ponding and ensure positive drainage to the desired surface inlet or drainage outlet.
- ❏ Typical full and partial sections through site are sufficiently detailed to show the relationship of finished floor elevation of building(s) to outside finished grades of both grassed and paved areas.
- ❏ The following typical sections are provided and adequately dimensioned:
  - (a) Concrete pavement
  - (b) Bituminous pavement
  - (c) Sidewalks, entrance drives, and roads
  - (d) Other sections, as required

- ❏ All applicable detail sketches and construction notes are shown for curb and gutter, storm drain inlets, manholes, headwalls, painting pavement markings, riprap, erosion control measures, and other required items of sitework. Appropriate specification sections are referenced when applicable.
- ❏ If the design includes concrete pavement, then the following must be shown:
  - (a) Concrete joint layout plan:
    1. Concrete joint details and spacing.
    2. Type of joint material, as per specifications.
    3. Special details for reinforced concrete slab around storm drain inlets, when required.
    4. Reinforcement of odd-shaped slabs.
    5. Tie-down anchors, as required.
    6. Other details, such as ADA requirements, as required.
- ❏ "Storm Drainage Pipe Structures Schedule" shown in the drawing detail(s) agrees with the drainage plan, drainage design analysis, and pipe profile(s) regarding inlet numbers, invert elevations, etc.
- ❏ Plant schedule agrees with the landscape plans.

❏ Locations of all soil borings, test pits, etc. are correctly shown on the Grading Plan, and appropriate symbols included in legend.

❏ All applicable detail sketches and construction notes are shown for erosion control measures, and other required items of sitework are finalized in the Erosion Control Plan. Appropriate specification sections are referenced when applicable.

❏ Project-specific details are essential. Generic "boiler-plate" terms are not adequate.

Storm Drainage Design

❏ Analysis contains an introductory page giving a brief description of the general terrain and/or site soil conditions, drainage patterns, basis of technical requirements, and other pertinent data affecting the proposed storm drainage system (formula, appropriate rainfall and runoff criteria, etc.).

❏ Drainage area map is complete, with subareas outlined, including possible "offsite" drainage, and all necessary "existing" and "new" drainage pipe and structures are indicated.

❏ Drainage tabulation forms are complete, and calculations are included for:

(a) ditch flow and culverts, when required.
(b) capacity and spacing of inlet openings.
(c) correct pipe strength(s) (Gauges/
   D-Loads).

Pavement Design Analysis

❑ Discussion of site conditions, etc. indicates
that borings logs have been reviewed to
ensure there are no unsuitable soils (heavy
clays/organic soils) that would require
removal and replacement in areas to be
paved or in other critical areas. If these conditions exist, then provisions have been
made for removal of same, and limits are
shown on the drawings.
❑ Classifications of road usage, vehicle category, CBR/K values, and method of deter-
mining required pavement thickness and
depths of compaction are satisfactory.
❑ The assumptions used in the pavement foundation analysis are consistent with the CBR
values specified in the final foundation report.

Landscape Design

❑ The sprinklers, lighting, hardscape, etc.
correspond with the site limits, including
the building and civil plans.

- ❏ Maintenance of landscape (watering, fertilizing, etc.) has been provided for in the design documents.
- ❏ Where applicable, appropriate "General Notes" are provided on the drawing(s), indicating trees to remain within the designated grading limits.
- ❏ All required plant items are included on plant list, and shrubs, etc. comply with approved plant list in the original Request for Proposal and/or other documents.
- ❏ Planting details (depths, size of hole, etc.) are provided.

CIVIL/SANITARY DESIGN CHECKLISTS

Sanitary Sewers

- ❏ Utility plan(s) show all existing and new sanitary sewers including manholes and cleanout locations.
- ❏ Sizes of sanitary sewers are shown, and all work can be located in the field from established benchmarks (BMs) or baselines.
- ❏ Sanitary sewers are profiled, including building connections, and show pertinent data (existing and final grades, top and invert elevations, size, length, pipe crossing).

- ❏ Building connections have been coordinated with interior plumbing size, inlet elevations, and locations.
- ❏ Sanitary sewers do not conflict with other underground utilities.
- ❏ Sewers are laid at sufficient slope to provide minimum velocity when flowing full.
- ❏ Minimum-size sewer lines are shown for building sewers and for mains.
- ❏ Adequate cover for frost protection has been provided.
- ❏ Determination made to maintain flow in existing sewer system during construction of new sewers.
- ❏ Abandoned sewers are shown as plugged or removed.
- ❏ Sanitary sewer appurtenance details are provided.

Water

- ❏ Pipe size is adequate for domestic water demand.
- ❏ Gate valves and valve boxes are properly located.
- ❏ Pipe size is adequate for fire flow demand.
- ❏ Number and location of new and existing fire hydrants are sufficient for adequate fire protection.

❑ Fire line entering building agrees with interior sprinkler plan. Position indicator valve (PIV) is shown.

Design Analysis

❑ Domestic water line(s) have been sized on correct fixture unit basis.
❑ Velocity and head loss have been computed.
❑ Sanitary and waterline specifications include all items, sizes, and work shown on the contract drawings. Inapplicable paragraphs indicated as "Not Used" and inapplicable reference publications have been deleted.
❑ All allowed pipe material options have been retained and correct strength of pipe has been selected.
❑ Special construction requirements are shown on details are properly covered and resolved in the specifications.

ARCHITECTURAL DESIGN CHECKLIST

❑ Site property lines and existing conditions match survey or civil drawings.
❑ Building location meets all setback requirements, zoning codes, and deed restrictions.
❑ Building limits match civil, plumbing, and electrical on-site plans.

- ❏ Locations of columns, bearing walls, grid lines, and overall building dimensions match structure.
- ❏ Locations of expansion joints—all floors—match with structural drawings.
- ❏ All elevated concrete slabs have a shoring and reshoring plan.
- ❏ Demolition instructions are clear on what to remove and what is to remain, and are coordinated with design documents.
- ❏ Building elevations match floor plans and have the same scale.
- ❏ Building sections match elevations, plans, and structural drawings.
- ❏ Building plan match lines are consistent on structural, mechanical, plumbing, and electrical drawings.
- ❏ Structural member locations are commensurate architecturally.
- ❏ Elevation points match structural drawings.
- ❏ Chases match structural, mechanical, plumbing, and electrical drawings.
- ❏ Section and detail call-outs are correct and cross-referenced.
- ❏ Large-scale plans and sections match small-scale plans and sections.
- ❏ Reflected architectural ceiling plans are coordinated with mechanical and electrical plans.

DESIGN

- ❏ Columns, beams, and slabs are listed on elevations and sections.
- ❏ Door schedule information matches plans, elevations, fire rating, and project manual.
- ❏ Cabinets or millwork will fit in available space.
- ❏ Flashing through the wall and weep holes are provided where code requires.
- ❏ Areas above halls and rooms are coordinated with mechanical, plumbing, and electrical plans (above ceiling cross section).
- ❏ Flashing materials and gauges are indicated or specified.
- ❏ Fire ratings of walls, ceilings, and fire and smoke dampers are indicated or specified.
- ❏ Adequate clearances have been given for the maintenance of all mechanical/electrical equipment as per code.
- ❏ Miscellaneous metals are detailed, noted, and coordinated with the Project Manual.
- ❏ Equipment room or areas are sized commensurate with mechanical, electrical, and plumbing equipment.
- ❏ Limits, types, and details of waterproofing are coordinated with design documents.
- ❏ Limits, types, and details of insulation are coordinated with design documents.

- ❏ Limits, types, and details of roofing are coordinated with design documents.
- ❏ Skylight structures are compatible with structural, mechanical, and electrical designs.
- ❏ Piping loads hang from the roof or floors, and are coordinated with the mechanical and structural drawings, and proper inserts are called for on the drawings.
- ❏ Mechanical and electrical equipment is properly supported, and all architectural features are adequately framed and connected.
- ❏ All drawings showing monorails, hoists, and similar items have support details, notes, and the locations are coordinated with the architectural, structural, mechanical, and electrical drawings.
- ❏ Walls, partitions, and window walls are not inadvertently loaded through deflection.
- ❏ All window walls, expansions, and weeps are provided.
- ❏ All physically disabled requirements are coordinated with plumbing and electrical plans.
- ❏ Architectural space requirements are commensurate with ductwork, conduit, piping, light fixtures, and other recesses.

- ❏ Architectural space requirements are commensurate with elevators, escalators, and other equipment.
- ❏ Dew point in walls, roof, and terraces, and vapor barrier have been provided as required.
- ❏ Concealed gutters are properly detailed, drained, and waterproofed; expansion has been provided for.
- ❏ Compatibility of grading around perimeter of building has been established with civil drawings.
- ❏ Color finish schedules are on drawings.
- ❏ Interior valleys for buildings having large flat roofs are provided with saddles or crickets to eliminate formation of "bird baths."
- ❏ Project-specific rather than generic "boiler plate" details are shown.

STRUCTURAL DESIGN CHECKLIST

- ❏ The design load conditions meet or exceed the Building codes and the Design Standards.
- ❏ The column orientation and grid lines on the structural and the architectural drawings match.
- ❏ The load-bearing walls and the building column foundation locations match with architectural drawings.

- ❏ The slab elevations match the architectural drawings.
- ❏ The depressed or raised slabs are indicated and match the architectural drawings.
- ❏ The limits of slabs on the structural drawings match the architectural drawings.
- ❏ The expansion joints on the structural drawings match the architectural drawings.
- ❏ The footing depth and cover are shown with the existing and final grades.
- ❏ The foundation piers, footings, and grade beams are coordinated with schedules.
- ❏ The footing and pier locations do not interfere with new and existing utilities, trenches, and tanks.
- ❏ The foundation wall elevations are the same as those on the architectural drawings.
- ❏ The location of door and roof framing column lines and column orientation match the foundation plan column lines and column orientation.
- ❏ The structural perimeter floor and roof lines match the architectural drawings.
- ❏ The section and detail call-outs are proper and cross-referenced.
- ❏ The columns, beams, and slabs are listed in schedules and are coordinated.

- ❏ The column length, beam, and joist depths match those same dimensions in the architectural drawings.
- ❏ The structural dimensions match the architectural drawings.
- ❏ The drawing notes do not conflict with specifications.
- ❏ The architectural construction and rustication joints are correct.
- ❏ The structural openings are coordinated with the architectural, mechanical, electrical, and plumbing drawings.
- ❏ The structural joist and beam locations do not interfere with water closets, floor urinals, floor drains, and chases.
- ❏ The structural design of roof and floors considered the superimposed loads, including the HVAC equipment, boilers, glass walls, etc.
- ❏ Cambers, drifts, and deflections have been coordinated with the architectural drawings.
- ❏ The concentrated load points on joists do not conflict with design by other disciplines, i.e., large water lines or fire main lines.
- ❏ Horizontal and vertical bracing, ladders, stairs, and framing do not interfere with doorways, piping, duct work, electrical, equipment, etc.

- ❑ The structural fire proofing requirements are coordinated with the architectural requirements.
- ❑ Rock excavation is a base bid or a unit price.
- ❑ Project-specific rather than generic "boiler plate" details are shown.

MECHANICAL DESIGN CHECKLISTS

Mechanical Design

- ❑ Mechanical plans match architectural and reflected ceiling plans.
- ❑ HVAC ducts are commensurate with architectural space and are not in conflict with conduit, piping, etc.
- ❑ Mechanical equipment fits architectural space with room for access, safety, and maintenance.
- ❑ Mechanical openings match architectural and structural drawings.
- ❑ Mechanical motor sizes match electrical schedules.
- ❑ Thermostat locations are not placed over dimmer controls.
- ❑ Equipment schedules correspond to manufacturer's specifications and design documents.

- ❑ Mechanical requirements for special equipment (i.e., kitchen, elevator, telephone, transformers) are clear.
- ❑ Fire damper located in ceiling and fire walls.
- ❑ All structural supports required for mechanical equipment are indicated on structural drawings.
- ❑ All roof penetrations are shown on roof plans.
- ❑ Seismic bracing details are provided for all platforms that support overhead equipment and seismic flexible coupling locations and details are shown.

Fire Protection Design

- ❑ Waterflow testing for all new sprinkler systems are conducted and waterflow test data indicated on drawings or in specifications.
- ❑ Detailed hydraulic calculations are provided verifying that water supply is sufficient to meet fire protection system demand.
- ❑ Complete riser diagram is shown.
- ❑ All piping from the point of connection to the top of the sprinkler riser(s) is shown on the drawings.

- ❑ All valves, fire department connections, and inspector's test connections are indicated on the drawings.
- ❑ Sprinkler main drain piping and discharge point are shown and detailed, and main drains discharge directly to the outside.
- ❑ The extent or limit of each type of sprinkler system, each design density, each type and temperature rating of sprinkler heads, and location of concealed piping is clearly specified or shown.
- ❑ Water-filled sprinkler piping is not subject to freezing.
- ❑ Detail of the sprinkler piping entry into the building is provided and includes details of anchoring and restraints.
- ❑ Aesthetics considerations are incorporated in the design of the sprinkler system; e.g., the sprinkler piping is concealed in finished areas and recessed chrome-plated pendent sprinkler heads are used in finished area.
- ❑ Paddle-type waterflow switches are only used in wet-pipe sprinkler systems. (The other sprinkler systems use pressure-type flow switches.)
- ❑ The main sprinkler control valves are accessible from the outside.
- ❑ Fire rating of fire-rated walls, partitions, floors, shafts, and doors are indicated.

- ❑ If spray-applied fire proofing is specified, the fire rating of the steel structural members is indicated.
- ❑ Locations of required fire dampers are shown.
- ❑ Location of all fire alarm indicating devices, pull stations, waterflow switches, detectors and other fire alarm and supervisory devices are indicated on the drawings.
- ❑ The connection of the fire alarm and detection system to the installation-wide fire alarm system is clearly shown and detailed.

Fire Alarm Plan (code required)

The following specs/dimensions will complete your Fire Alarm Plan:

- Floor plan
- Locations of alarm-initiating and notification equipment
- Alarm control and trouble signaling equipment
- Annunciation
- Power connection
- Battery calculations
- Conductor type and sizes
- Voltage drop calculations
- Manufacturers, model numbers, and listing information for equipment, devices, and materials

- Details of ceiling height and construction.
- The interface of fire safety control functions.

Plumbing Design

❑ Plumbing plans match architectural, mechanical, and structural drawings.
❑ Plumbing fixtures match plumbing schedules and architectural locations.
❑ Compatibility of site piping limits interfaces with building piping.
❑ Roof drain locations coordinate with roof plan.
❑ Subsurface drains are located and detailed.
❑ Roof drain overflows are provided.
❑ Piping chase locations match architectural and structural drawings.
❑ All hot and cold water piping is insulated IAW, the contractor's approved piping insulation display sample.
❑ Piping is commensurate with architectural space and not in conflict with conduit, ducts, and structure elements.
❑ Piping openings match architectural and structural drawings.
❑ Complete riser plans are shown with *all* piping!
❑ Structural design is compatible with plumbing equipment and piping requirements.

- ❏ Plumbing equipment schedules correspond to manufacturers' specifications and design documents.
- ❏ Floor drains match architectural and kitchen equipment plans.
- ❏ Site utilities have been accurately verified, and site water and gas service requirements are met by supply utilities.
- ❏ Floor openings, i.e., drains, water closets, do not conflict with structural beams, joists, or trusses.
- ❏ Limits and confines where piping may be run are shown.
- ❏ Seismic bracing details are provided, and seismic flexible coupling locations are shown.
- ❏ Roof drain details are coordinated with other trades to show the installation of sump pans in ribbed sheet metal decks, and the placement of roof insulation in and around the drainage fitting.

Electrical Design

- ❏ Electrical plans match architectural, mechanical, plumbing, and structural plans.
- ❏ Location of light fixtures, speakers, etc. match with reflected ceiling plans.

- ❏ Electrical connections are shown for equipment, e.g., mechanical motors, heat strips, architectural, overhead doors, stoves, dishwashers, etc.
- ❏ Locations of panel boards and transformers are shown on architectural, mechanical, and plumbing plans.
- ❏ Conduit chase locations match with architectural and structural drawings.
- ❏ Compatibility of conduit and light fixtures with architectural space is met; no conflicts exist with duct, piping, or structure.
- ❏ Electrical equipment structural requirements have been met.
- ❏ Electrical equipment room fits architectural space, with clearance for safety and maintenance.
- ❏ Electrical voltage, phasing for all motors match on mechanical and architectural designs.
- ❏ Fixtures, speakers, clocks, etc. schedules correspond to a manufacturer's description and design documents.
- ❏ Light fixture spacing and location are arranged and specified to prevent dark spots.
- ❏ Location of duplex outlets, telephone, fire alarms, clock outlets, etc. with architectural millwork and finishes as indicated on plan.

- ❏ Limits and confines where conduits may be run have been verified.
- ❏ Site electrical and telephone service requirements have been verified with the supply utility.
- ❏ Seismic bracing details have been provided, and seismic flexible coupling locations have been shown.

DRAWING CHECKLIST

- ❏ All work depicted on drawings is readable at full size.
- ❏ New work is shown three pen weights heavier than existing construction.
- ❏ Overlays and base sheet are composited to check for duplication or overprinting of features, notes, plans, sections, and details.
- ❏ Titles, subtitles, scales, title block, and revision block information is complete and accurate.
- ❏ Titles of drawings agree with the titles listed on the Index of Drawings.
- ❏ The total number of drawings is on the first sheet of the set and is correct.
- ❏ The signature block is on the first sheet of each discipline.
- ❏ Drawings are consecutively numbered.
- ❏ All drawings are present.

- ❏ Site-adapted drawings have the appropriate notation in each revision block.
- ❏ Amended or modified drawings have the appropriate notation in each revision block.
- ❏ Symbols on drawings are standard and accompanied by the complete legend.
- ❏ The use of cross-referencing bubbles for locating sections, details, and elevations has been coordinated and explained.
- ❏ On the Final Design submittal, all title block numbering (Plate No., File No., Sheet No., and Ring No.) is in place.
- ❏ All final contract drawings are free of tape, appliques, and shading.
- ❏ Colored ink is not used on plotted drawings.
- ❏ Multiple drawing layers have been composited into either a single reproducible sheet or into one reproducible sheet per color overlay where color reproduction is planned.

SPECIFICATION CHECKLIST

- ❏ Project name, location, and project number are inserted as a main heading at the top of each subheading on the first page of each section.
- ❏ All "gaps" have been eliminated where material has been omitted from text.

- ❑ Other technical section(s) referenced within a section have been included, and either the section has been added or the paragraph rewritten to eliminate the reference.
- ❑ Omitted main paragraphs indicated as "NOT USED," and omitted subparagraphs indicated as "Not used."
- ❑ Consecutively omitted paragraphs are single spaced.
- ❑ All blanks have been filled in and all brackets removed.
- ❑ All tables have been printed on one page (unless it is physically impossible to fit the table on one page).
- ❑ If tables require more than one page, headings have been duplicated on second page.
- ❑ Margins are properly set a minimum of 25 mm (1 in) on all four sides of the sheet (right, left, top, and bottom).
- ❑ Page numbers are shown at least 12 mm (½ in) from the bottom of the page and prefixed with the section number.
- ❑ Page numbers are correct.
- ❑ Paragraphs numbers are connected.
- ❑ Submittal Register has been verified with the owner.
- ❑ Verify that all required sections of the project specifications have been included.

- ❑ Verify that the appropriate review level has been indicated for all submittals listed on the Contractor Submittal Register, and that the Register agrees with the technical specification sections.
- ❑ Require all engineering disciplines to review and sign off on a legal document that they have fully read all specifications, plans, and construction documents, and that to the best of their knowledge contains no errors or omissions and that adequate quality review has been completed. All reviews have been signed off by the Design Project Manager.

IBC CODE REQUIREMENTS

TABLE 2-1 BUILDING CLASSIFICATIONS	
Assembly	Group A
Business	Group B
Educational	Group E
Factory	Group F
High-hazard	Group H
Institutional	Group I
Mercantile	Group M
Residential	Group R
Storage	Group S
Utility and miscellaneous	Group U

TABLE 2-2 REQUIRED RATED WALL SEPARATIONS

Room or Area	Separation
Furnace room where any piece of equipment is over 400,000 Btu per hour input	1 hour or provide automatic fire-extinguishing system
Rooms with any boiler over 15 psi and 10 horsepower	1 hour or provide automatic fire-extinguishing system
Refrigerant machinery rooms	1 hour or provide automatic sprinkler system
Parking garage	2 hours, or 1 hour and provide automatic fire-extinguishing system
Hydrogen cut-off rooms	1-hour fire barriers and floor/ceiling assemblies in Group B, F, H, M, S, and U occupancies. 2-hour fire barriers and floor/ceiling assemblies in Group A, E, I, and R occupancies.
Incinerator rooms	2 hours and automatic sprinkler system
Paint shops, not classified as Group H, located in occupancies other than Group F	2 hours, or 1 hour and provide automatic fire-extinguishing system
Laboratories and vocational shops, not classified as Group H, located in Group E or 1–2 occupancies	1 hour or provide automatic fire-extinguishing system
Laundry rooms over 100 sq ft	1 hour or provide automatic fire-extinguishing system
Storage rooms over 100 sq ft	1 hour or provide automatic fire-extinguishing system
Group I: 3 cells equipped with padded surfaces	1 hour

Room or Area	Separation
Group I: 2 waste and linen collection rooms	1 hour
Waste and linen collection rooms over 100 sq ft	1 hour, or provide automatic fire-extinguishing system
Stationary lead-acid battery systems having a liquid capacity of more than 100 gallons used for facility standby power, emergency power, or uninterrupted power supplies	1-hour fire barriers and floor/ceiling assemblies in Group B, F, H, M, S, and U occupancies. 2-hour fire barriers and floor/ceiling assemblies in Group A, E, I, and R occupancies

Where an automatic fire-extinguishing system is provided, it need only be provided in the incidental use room or area.

TABLE 2-3   MINIMUM CODE REQUIRED DESIGN LOADING

Occupancy/use	Uniform Load (psi)	Concentrated Load (lb)
Apartments (see residential)	—	—
Access floor systems		
Office use	50	2000
Computer use	100	2000
Armories and drill rooms	150	—
Assembly areas and theaters		
Fixed seats (fastened to floor)	60	
Lobbies	100	
Movable seats	100	
Stages and platforms	125	—
Projections, and control rooms	50	
Catwalks	40	
Balconies (exterior)	100	
On one- and two-family residences only, and not exceeding 100 ft²	60	—
Decks	Same as occupancy served	
Bowling alleys	75	—
Cornices	60	—
Corridors, except as otherwise indicated	100	—
Dance halls and ballrooms	100	—
Dining rooms and restaurants	100	—
Elevator machine room grating (on area of 4 in²)	—	300
Finish light floor plate construction (on area of 1 in²)		200

Occupancy/use	Uniform Load (psi)	Concentrated Load (lb)
Fire escapes	100	
On single-family dwellings only	40	—
Garages (passenger vehicles only)	40	
Grandstands (see stadium and arena bleachers)	—	—
Gymnasiums, main floors, and balconies	100	—
Handrails, guards, and grab bars	See Section 1607.7 of IBC	
Hospitals		
Operating rooms, laboratories	60	1000
Private rooms	40	1000
Wards	40	1000
Corridors above first floor	80	1000
Hotels (see residential)	—	—
Libraries		
Reading rooms	60	1000
Stack rooms	150	1000
Corridors above first floor	80	1000
Manufacturing		
Light	125	2000
Heavy	250	3000
Marquees	75	—
Office buildings		
File and computer rooms will be designed for heavier loads based on anticipated occupancy.		
Lobbies and first-floor corridors	100	2000
Offices	50	2000
Corridors above first floor	80	2000
		(continued)

DESIGN

TABLE 2-3 (CONTINUED)

Occupancy/use	Uniform Load (psi)	Concentrated Load (lb)
Penal institutions		
Cell blocks	40	—
Corridors	100	
Residential		
One- and two-family dwellings		
Uninhabitable attics without storage	10	
Uninhabitable attics with storage	20	
Habitable attics and sleeping areas	30	
All other areas except balconies and decks	40	
Hotels and multifamily dwellings		
Private rooms and corridors serving them	40	
Public rooms and corridors serving them	100	
Schools		
Classrooms	40	1000
Corridors above first floor	80	1000
First-floor corridors	100	1000
Scuttles, skylight ribs, and accessible ceilings	—	200
Sidewalks, vehicular driveways, and yards subject to trucks	250	8000
Skating rinks	100	—
Stadiums and arenas		
Bleachers	100	—
Fixed seats (fastened to floor)	60	
Stairs and exits	100	
One- and two-family dwellings	40	
All other	100	

Occupancy/use	Uniform Load (psi)	Concentrated Load (lb)
Storage warehouses (shall be designed for heavier loads if required for anticipated storage)		
Light	125	
Heavy	250	
Stores		
Retail		
First floor	100	1000
Upper floors	75	1000
Wholesale, all floors	125	1000
Walkways and elevated platforms (other than exit ways)	60	—

TABLE 2-4   MAXIMUM FLOOR AREA ALLOWANCES PER OCCUPANT

Occupancy	Floor Area in Square Feet per Occupant
Agricultural building	300 gross
Aircraft hangars	500 gross
Airport terminal	
Concourse	100 gross
Waiting areas	15 gross
Baggage claim	20 gross
Baggage handling	300 gross
Assembly	
Gaming floors (keno, slots, etc.)	11 gross
Assembly with fixed seats	See 1003.2.2.9
Assembly without fixed seats	
Concentrated (chairs only—not fixed)	7 net
Standing space	5 net
Unconcentrated (tables and chairs)	15 net
Bowling centers: allow 5 persons for each lane, including 15 ft of runway, and for additional areas	7 net
Business areas	100 gross
Courtrooms—other than fixed seating areas	40 net
Dormitories	50 gross
Educational	
Classroom area	20 net
Shops and other vocational room areas	50 net
Exercise rooms	50 gross

Occupancy	Floor Area in Square Feet per Occupant
H-5 Fabrication and manufacturing areas	200 gross
Industrial areas	100 gross

Institutional areas
  Inpatient treatment areas
  Outpatient areas
  Sleeping areas
	
	240 gross
	100 gross
	120 gross

Kitchens: commercial	200 gross

Library
  Reading rooms
  Stack area
	
	50 net
	100 gross

Locker rooms	50 gross

Mercantile
  Basement and grade floor areas
  Areas on other floors
  Storage, stock
	
	30 gross
	60 gross
	300 gross

Parking garages	200 gross
Residential	200 gross

Skating rinks, swimming pools
  Rink and pool
  Decks
	
	50 gross
	15 gross

Stages and platforms	15 net
Accessory storage areas, mechanical equipment room	300 gross
Warehouses	500 gross

TABLE **2-5**	MEANS OF EGRESS	
Occupancy	**Without Sprinkler System (ft)**	**With Sprinkler System (ft)**
A, E, F-I, I-I, M, R, S-I	200	250
B	200	300
F2, S-2, U	300	400
H-I	Not permitted	75
H-2	Not permitted	100
H-3	Not permitted	150
H-4	Not permitted	175
H-5	Not permitted	2000
1–2, 1–3, 1–4	150	200

See IBC Code and IFC Codes for special circumstances.

Common path of egress travel. In occupancies other than Groups H-I, H-2, and H-3, the common path of egress travel shall not exceed 75 ft (22 860 mm).

In occupancies in Groups H-I, H-2, and H-3, the common path of egress travel shall not exceed 25 ft (7620 mm).

Exceptions:

1. The length of a common path of egress travel in an occupancy in Groups B, F, and S shall not be more than 100 ft provided that the building is equipped throughout with an automatic sprinkler system installed.

2. Where a tenant space in an occupancy in Group B, S and U has an occupant load of not more than 30, the length of a common path of egress travel shall not be more than 100 ft. The length of a common path of egress travel in occupancies in Group 1–3 shall not be more than 100 ft

TABLE 2-6 CODE REQUIRED NUMBER AND DISTRIBUTION OF FIRE HYDRANTS

Fireflow Requirement (gpm)	Minimum Number of Hydrants	Average Spacing between Hydrants (ft)	Maximum Distance from Any Point on Street or Road Frontage to a Hydrant
1,750 or less	1	500	250
2,000–2,250	2	450	225
2,500	3	450	225
3,000	3	400	225
3,500–4,000	4	350	210
4,500–5,000	5	300	180
5,500	6	300	180
6,000	6	250	150
6,500–7,000	7	250	150
7,500	8	200	120

- Reduce by 100 ft for dead-end streets or roads.
- Where streets are provided with median dividers that can be crossed by firefighters pulling hose lines, or where arterial streets are provided with four or more traffic lanes and have a traffic count of more than 30,000 vehicles per day, hydrant spacing shall average 500 ft on each side of the street and be arranged on an alternating basis up to a fire-flow requirement of 7000 gallons per minute and 400 ft for higher fire-flow requirements.
- Where new water mains are extended along streets where hydrants are not needed for protection of structures or similar fire problems, fire hydrants shall be provided at spacing not to exceed 1000 ft to provide for transportation hazards.
- Reduce by 50 ft for dead-end streets or roads.
- One hydrant for each 1000 gallons per minute or fraction thereof.

DESIGN

TABLE 2-7 TYPES OF CONSTRUCTION

Type I	Fire-resistive construction. This is the most fire-resistant construction. Frames are 3-hour rated
Type II	Fire-resistive with 1 hour, or nonprotected construction. Frames either have a 2–1 hour or no fire-resistive protection
Type III	1 hour, or nonprotected construction. Frames are conbustible with 1 hour or no fire-resistive protection
Type IV	Heavy timber construction. Frames are combustible but slow to burn
Type V	1 hour or nonprotected. Framed with light wood framing

TABLE 2-8		FIRE SEPARATIONS (HOURS) FOR MIXED USE OCCUPANCIES							
	A-1	A-2	A-3	A-4	A-5	B	E	F-1	F-2	H
A-1	NA	3	2	2	2	2	2	2	2	4
A-2		3	3	3	3	3	3	3	3	4
A-3			2	2	2	2	2	2	2	4
A-4				NA	2	2	2	2	2	4
A-5					NA	2	2	2	2	4
B						NA		22	2	4
E							NA	2	2	4
F-1								2	2	4
F-2									NA	4
H										NA
I-1										
I-2										
I-3										
M										
R-1										
R-2										
R-3										
S-1										
S-2										
U										

TABLE 2-9 FIRE SEPARATIONS (HOURS) FOR MIXED USE OCCUPANCIES (CONTINUED)

	I-1	I-2	I-3	M	R-1	R-2	R-3	S-1	S-2	U
A-1	2	3	3	2	2	2	2	2	2	NA
A-2	3	3	3	3	3	3	3	3	3	NA
A-3	2	3	3	2	2	2	2	2	2	NA
A-4	2	3	3	2	2	2	2	2	2	NA
A-5	2	3	3	2	2	2	2	2	2	NA
B	2	3	3	2	2	2	2	2	2	NA
E	2	3	3	2	2	2	2	2	2	NA
F-1	2	3	3	2	2	2	2	2	2	NA
F-2	2	3	3	2	2	2	2	2	2	NA
H	4	4	4	4	4	4	4	4	4	NA
I-1	NA	3	3	2	2	2	2	2	2	NA
I-2		NA	3	3	3	3	3	3	3	NA
I-3			NA	3	3	3	3	3	3	NA
M				2	2	2	2	2	2	NA
R-1					NA	2	2	2	2	NA
R-2						NA	2	2	2	NA
R-3							NA	2	2	NA
S-1								2	2	NA
S-2									NA	NA
U										NA

CHAPTER 3
DUCTWORK

GENERAL

This chapter covers ductwork for air conditioning, heating, ventilating, and exhaust systems. It is important that the Inspector have a thorough knowledge of the job plans, specifications, and potential obstructions in the area in which the ductwork is to be installed, including locations of fire rated walls that the duct must penetrate.

The checklists in this chapter are meant to serve as guides to ensure that regulations as applied to ductwork are met.

❏ All equipment has identification nameplates, and the unit is as specified.
❏ Approved vibration isolators and flexible connections are furnished and installed if required.
❏ Using building equipment for temporary heat is understood and/or approved.
❏ Provisions are made for proper mounting and anchorage of equipment pads, hangers, etc. Special code requirements for seismic zone areas.
❏ Equipment operates as intended.

SHOP DRAWINGS

- ❑ It is the Project Architect/Engineer's responsibility to determine that all ductwork is approved well in advance to avoid progress delays.
- ❑ Check ductwork delivered to the site for conformance with approved shop drawings.
- ❑ Make sure delivered and stored items are properly stored with tags, so that they will not be installed in a wrong location.

Fabrication*

- ❑ Inspect for type, thickness, and shape of sheet material, and fiberglass boards used for ductwork.
- ❑ Check workmanship and observe lock seams and breaks in ductwork for cracks of sheet metal ducts. Check fiberglass ducts for broken, or damaged edges, joints, and seams.
- ❑ Inspect all joint connections for correct type and be certain they are adequately sealed to prevent movement and air loss.

---

*See *SMACNA Duct Manual* appropriate to material and service requirements.

- ❏ Make sure that the joints are neatly finished and that the duct is smooth on the inside. All laps should be made in the directions of air flow. Internal insulation will be securely fastened and coated as specified.
- ❏ Provide adequate bracing and reinforcement of the ducts.
- ❏ Compare the radius of curved duct with the specification requirements.
- ❏ Check the slope ratio of all transitions. Provide turning vanes and extractors to eliminate abrupt turns of air which cause noticeable turbulence (oil canning).
- ❏ Check the need for and construction of splitter dampers. Make sure the operating mechanism is accessible; if exposed in a room, the mechanism must be finished.
- ❏ Make sure that fire and/or smoke dampers are provided in ducts as required in accord with NEPA and SMACNA fire damper book and codes.
- ❏ Check for fire-safety switch on return air ducts of circulation system.
- ❏ Check duct for the required test holes and covers. Are they accessible?
- ❏ Make sure that the ducts are sealed and protected during the construction period
- ❏ Check the fabrication of flexible connections.

DUCT-WORK

- ❏ All equipment serviced by the ductwork is required to be fully accessible for maintenance, repairs, oiling, cleaning, and filter changing.

INSTALLATION

- ❏ Examine all fabricated ducts, rejecting any which are not smooth or any which are damaged.
- ❏ Examine duct hangers for specified material, thickness, and spacing.
- ❏ Check specification requirements for the need for stiffeners for wide ducts.
- ❏ Check for need of trapeze hangers under wide ducts.
- ❏ Provide approved flexible connections between ducts and for fan units.
- ❏ Check rigidity and tightness of field-installed items as dampers and defectors.
- ❏ Provide access doors at all fire dampers, automatic dampers, coils, filters, heaters, thermostats, or at any item that requires servicing. Doors are to be airtight, securely fastened and accessible, and able to be fully opened. Refer to SMACNA and specifications for required size of access doors.
- ❏ Inspect goose necks and rain hoods for method of fastening, flashing, and bracing.

Goose necks are to be turned away from the prevailing wind. Check specifications for screens on the open end of the goose necks.

❑ Provide properly sized sleeves where insulated duct passes through wall openings. Future requirements?

❑ When obstructions cannot be avoided, the duct area should never be decreased more than 10%, and then a streamlined collar should be used. Larger obstructions require an increase in the duct size in order to maintain as nearly uniform velocity as possible.

Note: Test metal duct for air tightness before insulating.

All ducts, plenums, and casings must be thoroughly cleaned of debris and blown free of small particles and dust before supply outlets are installed.

THINK SAFETY AT ALL TIMES

CHAPTER 4
INSTALLATION

DRILLING AND NOTCHING FOR WOOD CONSTRUCTION

Studs

- LOAD BEARING
  (a) Holes less than 40%
  (b) Notches less than 25%
- NONLOAD BEARING
  (a) Holes less than 60%
  (b) Notches less than 40%

Notching and Boring Studs

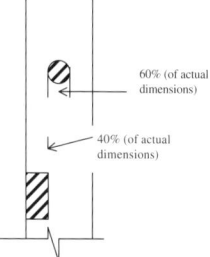

60% (of actual dimensions)

40% (of actual dimensions)

Figure 4-1   LOAD-BEARING STUD LIMITATIONS

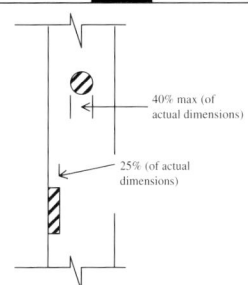

40% max (of actual dimensions)

25% (of actual dimensions)

Figure 4-2 NONLOAD-BEARING STUD LIMITATIONS

TABLE 4-1 HOLES/NOTCHING IN LOAD-BEARING STUDS

Nominal	Actual	Holes		Notching
		40%	60%	25%
2 × 4	1½ × 3½	1¹³⁄₃₂	2³⁄₃₂	⅞
2 × 6	1½ × 5½	2³⁄₁₆	3⁵⁄₁₆	1⅜

TABLE 4-2 HOLES/NOTCHING IN NONLOAD-BEARING STUDS

Nominal	Actual	Holes		Notching
		40%	60%	40%
2 × 4	1½ × 3½	1¹³⁄₃₂	2³⁄₃₂	1¹³⁄₃₂
2 × 6	1½ × 5½	2³⁄₁₆	3⁵⁄₁₆	2³⁄₁₆

INSTALL

Joists

- Notches less than 1/6 the depth
- **No notches or cuts are allowed in middle third!**
- Not within 2 in. of top or bottom
- Drilled holes less than 1/3 the depth
- Holes must be minimum of 2 in. from notches

TABLE 4-3	DRILLED HOLES MAXIMUMS IN JOISTS	
Nominal	Actual	1/3 (33%) in.
2 × 6	1½ × 5¼	1¾
2 × 8	1½ × 7¼	2¹³⁄₃₂
2 × 10	1½ × 9¼	3³⁄₃₂
2 × 12	1½ × 11¼	3¾
2 × 14	1½ × 13¼	4¾

TABLE 4-4	NOTCH MAXIMUMS IN JOISTS	
Nominal	Actual	1/6 (17%) in.
2 × 6	1½ × 5¼	⅞
2 × 8	1½ × 7¼	1⁷⁄₃₂
2 × 10	1½ × 9¼	1¹⁷⁄₃₂
2 × 12	1½ × 11¼	1⅞
2 × 14	1½ × 13¼	2⅜

TABLE 4-5	NOTCH MAXIMUMS IN CEILING JOISTS/RAFTERS	
Nominal	**Actual**	**$\frac{1}{3}$ (33%) in.**
2×6	$1\frac{1}{2} \times 5\frac{1}{4}$	$1\frac{3}{4}$
2×8	$1\frac{1}{2} \times 7\frac{1}{4}$	$2\frac{13}{32}$
2×10	$1\frac{1}{2} \times 9\frac{1}{4}$	$3\frac{3}{32}$
2×12	$1\frac{1}{2} \times 11\frac{1}{4}$	$3\frac{3}{4}$
2×14	$1\frac{1}{2} \times 13\frac{1}{4}$	$4\frac{3}{4}$

TABLE 4-6	MAXIMUMS END NOTCH IN CEILING JOISTS/RAFTERS	
Nominal	**Actual**	**$\frac{1}{4}$ (25%) in.**
2×6	$1\frac{1}{2} \times 5\frac{1}{4}$	$1\frac{5}{16}$
2×8	$1\frac{1}{2} \times 7\frac{1}{4}$	$1\frac{13}{16}$
2×10	$1\frac{1}{2} \times 9\frac{1}{4}$	$2\frac{5}{16}$
2×12	$1\frac{1}{2} \times 11\frac{1}{4}$	$2\frac{13}{16}$
2×14	$1\frac{1}{2} \times 13\frac{1}{4}$	$3\frac{5}{16}$

INSTALL

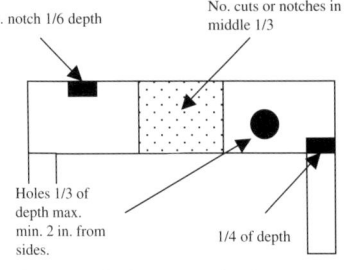

Max. notch 1/6 depth

No. cuts or notches in middle 1/3

Holes 1/3 of depth max. min. 2 in. from sides.

1/4 of depth

Note: See Up to Code Framing for Cutting, Drilling, or Notching Engineered Joist.

Figure 4-3   NOTCHES AND HOLES SPACING AND SIZES ALLOWED FOR JOIST

DIFFUSERS, REGISTERS, AND GRILLES

❏ Ensure that the contractor furnishes a schedule showing all air inlets and outlets.
❏ Inspect diffusers and registers for accessible volume control operator.
❏ Examine specification and installation for integral antismudge rings for diffusers.
❏ Check for loose or bent vanes.
❏ Inspect each item for fit, and see that gaskets are provided when required.
❏ Inspect for the proper operation of registers, dampers, and grille vent controls.

DUCTWORK INSTALLATION

❑ Ductwork layout must be coordinated with other trades to avoid congestion and interference. A ductwork drawing coordinating plumbing, electrical, sprinklers, etc. is recommended on complex work.

❑ Be sure type, material, thickness, and shape are as required.

❑ Joint connections must be required type. Check seams and breaks for cracks. Be sure the joint provides a smooth surface on interior of duct, and that laps are in direction of air flow.

❑ Slope ratio of transitions, radius of curved duct, air turns, and deflectors are provided as required.

❑ Bracing, reinforcement, stiffeners, hangers, etc. must be provided and ductwork must be installed as plans and specifications.

❑ Verify that all volume dampers, branch duct dampers, register or diffuser dampers, and splitter dampers are provided as required and operating mechanism is accessible.

❑ Fire dampers and smoke dampers of type required must be furnished and installed as required by NFPA. Verify that access is provided to dampers.

- ❏ Flexible connectors must be fabricated and provided where required.
- ❏ Access doors and/or access space must be provided for all items requiring servicing, such as fire dampers, automatic dampers, manual dampers, coils, heaters, filters, and thermostats. Size must be sufficient for access and maintenance.
- ❏ Proper sleeves and openings through walls and floors must be provided and sealed as required. *Allow no cutting of structural members without approval.*
- ❏ Ductwork must be properly taped or sealed as required by the codes and the contract specifications.
- ❏ Plastic duct may not exceed 150 degrees.
- ❏ All ductwork in nonconditioned areas needs to be insulated.
- ❏ Cooling ducts passing through nonconditioned areas also need a vapor retarder.
- ❏ All ducts, plenums, and equipment must be thoroughly cleaned of all debris before supply outlets are installed.
- ❏ All items must be furnished and installed as required, and approved.
- ❏ Finishes in areas must match as required.
- ❏ Volume control devices must be provided and accessible as required.

❏ Gaskets must be provided and installed as required.
❏ Items must be securely attached and supported as required.

Supports

❏ Metal ducts to be supported must be 1-in. straps of 18-gage or 12-gage galvanized metal wire.
❏ _Supports must not exceed 10-ft intervals._

THINK SAFETY AT ALL TIMES

CHAPTER 5
INSULATIONS

GENERAL

This chapter covers field-applied insulation. Factory-applied insulation is specified under the equipment, duct, or piping to be installed, as detailed in the specifications and plans.

IDENTIFICATION OF MATERIAL

All packages or standard containers of insulation, jacket material, cements, adhesives, and coatings delivered for use, and all samples must have a manufacturer's stamp or label attached giving the name of the manufacturer and brand, and a description of the material.

SHOP DRAWINGS

It is the inspector's responsibility to determine that all insulation-related materials are approved well in advance of their actual need on the job.

After approval of materials and prior to insulating any pipe, the contractor will submit for approval sample insulation boards, or approved standards, showing his or her proposed methods of mechanical insulation, including cut-a-way sections, insulation, coverings, and finish of completed work. Approved sample boards must be maintained by the contractor at the jobsite for the duration of the work.

SURFACE BURNING CHARACTERISTICS

❏ Check underwriters' labels and test certificate of all insulating materials and accessories for not exceeding a flame spread rating of 25 or a smoke-developed rating of 50, as determined by ASTM E 84.
❏ Check specification for limitations on surface burning characteristic.

Material Classes

CLASS A: Flame spread 0–25
Smoke-developed index 0–450

CLASS B: Flame spread 26–75
Smoke-developed index 0–450

CLASS C: Flame spread 76–200
Smoke-developed index 0–450

TABLE 5-1 INTERIOR WALL AND CEILING FINISH REQUIREMENTS WHEN SPRINKLERED

GROUP	Sprinklers		
	Vertical Exits and Exit Passageways	Exit Access Corridors and Other Exit Ways	Rooms and Enclosed Space
A-II& A-2	B	B	C
A-3f, I A-4, A-5	B	B	C
B,E,M,R-I,R-4	B	C	C
F	C	C	C
H	B	B	0
I-I	B	C	C
1-2	B	B	B
1-3	A	A	C
1-4	B	B	B
R-2	C	C	C
R-3	C	C	C
S	C	C	C
U	No restrictions		

TABLE 5-2 INTERIOR WALL AND CEILING FINISH REQUIREMENTS WHEN NOT SPRINKLERED

GROUP	No Sprinklers		
	Vertical Exits and Exit Passageways	Exit Access Corridors and Other Exit Ways	Rooms and Enclosed Space
A-II& A-2	A	A	B
A-3f, I A-4, A-5	A	A	C
B,E,M,R-I,R-4	A	B	C
F	B	C	C
H	A	A	B
I-1	A	B	B
1-2	A	A	B
1-3	A	A	B
1-4	A	A	B
R-2	B	B	C
R-3	C	C	C
S	B	B	C
U	No restrictions		

DUCTWORK INSULATION

❏ Distinguish between areas requiring flexible type insulation and those requiring rigid or semirigid type insulation.
❏ Check the type and thickness of insulation and requirements for vapor barrier.

INSULATE

- ❑ Check the method of fastening insulation to exterior or interior of duct.
  - (a) If metal pins are used, check the type and spacing.
  - (b) If wire is used, see that corners of insulation are protected from possible damage.
  - (c) Verify that adhesive materials are correct, and that the area specified receives proper coverage.
- ❑ Make a careful check for breaks in insulation and vapor barriers.
- ❑ See that materials are fire-retardant or non-combustible as required by the specifications.
- ❑ When equipment casings are required to be insulated, check for proper, firm application.
- ❑ Where insulation is subject to mechanical damage, check for protection requirements.
- ❑ Check for continuity of insulation through walls and floor, if required.
- ❑ Check for proper sealing of insulation to diffusers, grilles, and fire dampers.

DUCT INSULATION

Be sure the following steps are completed to ensure proper duct insulation.

- ❑ Ducts are tested for air tightness, if required, before installation of insulation.

❑ Type, thickness, material, extent, and method of fastening and installation are as required.

❑ Sound deadening and vapor barrier are provided as required.

❑ Insulation subject to damage is protected as required.

❑ Materials are fire retardant or incombustible as required.

❑ Vapor barrier integrity is maintained.

Duct Mastic

Duct mastic is a preferred flexible sealant that can move with the expansion, contraction, and vibration of the duct system components. It is often strengthened with fiberglass strands for increased strength. However, mastic is not a substitute for mechanical fastening of duct system components.

Note the following guidelines for correct usage and application of duct mastic.

• Choose water-based mastic, which is the least toxic and easiest to clean up.

• If a gap exceeds ¼ in., reinforce mastic with fiberglass mesh tape.

• Conventional duct tape should not be used in a duct system, except to seal the joints on access doors.

- The application process for mastic requires that all duct connections be mechanically fastened with screws, rivets, or, when using flex duct, with metal bands.
- The area to be joined should be wiped clean with a dry rag.
- The mastic is then applied with a trowel or brush (according to its viscosity) and spread 1 in. beyond the opening.
- For ¼- to ½-in. openings, use fiberglass mesh tape under the mastic.
- Gaps larger than ¼ to ½ in. require a rigid material covering.
- All connections (splices, Y's, T's, and boots) must be sealed.
- All boots must be sealed to the sheetrock (a wire can be used to keep it from pulling loose).
- All penetrations into a plenum must be sealed.
- Flex duct inner and outer linings need to be sealed (do not extend the duct liner through the wall of the plenum to the interior of the plenum).
- The air handler closet and air handler itself must be sealed, including sealing the air handler to the platform.
- The return plenum should be lined on the interior with duct board (foil face in) and

sealed. The support platform should be sealed on all sides.

- Penetrations into the plenum, such as refrigerant lines, must be sealed.
- Return air grills should be sealed at the point of wall penetration.
- Seal boots to sheetrock with approved materials.

PIPE INSULATION

❑ Determine whether the material on the job has been approved for the particular piping being installed. Make sure insulations, vapor barriers, adhesives, and sealers are noncombustible or fire retardant as specified. (Heated water piping is insulated differently from chilled water piping and from combination chilled and heated water piping.)

❑ Check thickness of insulation and of vapor barrier.

❑ Determine if insulation jackets which are exposed to view are required to be painted.

❑ Examine the requirements for the insulation of flanges, fittings, and valves, and ensure compliance with the requirements and specifications of the project.

- ❑ Check the lap and the sealing at joints.
- ❑ Be very careful to see that there are no breaks in the vapor barrier. Watch for later damages during construction.
- ❑ Check specification requirements for extending through sleeves in walls, floors, and ceilings; chilled water lines inside cabinets of fan coil units should be covered as required to prevent condensate dripping on floor.
- ❑ Make sure that pipe hangers are installed over insulation. Metal shields must be provided between hanger ring and insulation. High-density insulation inserts must be installed with a length equal to the length of metal shield.
- ❑ Check for the neat termination and seal of insulation at the end of insulation.
- ❑ Know the special requirements for insulation and jacketing of piping exposed to weather.
- ❑ Check the installation, width, and spacing of the bands used on pipe jacketing.
- ❑ In chilled-water and hot-water combination piping, check for vapor seal requirement on boiler piping.

TABLE 5-3   MINIMUM PIPE INSULATION THICKNESS

Fluid Design Operating Temp. Range (°F)	Mean Rating Temp. (°F)	Nominal Pipe or Tube Size (in.)				
		<1	1 to <1½	1½ to < 4	4 to <8	>= 8
Heating Systems (Steam, Steam Condensate and Hot Water)						
>350	250	2.5	3.0	3.0	4.0	4.0
251–350	200	1.5	2.5	3.0	3.0	3.0
201–250	150	1.5	1.5	2.0	2.0	2.0
141–200	125	1.0	1.0	1.0	1.5	1.5
105–140	100	0.5	0.5	1.0	1.0	1.0
Domestic and Service Hot Water Systems						
105 and Greater	100	0.5	0.5	1.0	1.0	1.0
Cooling Systems (Chilled Water, Brine, and Refrigerant)						
40–60	100	0.5	0.5	1.0	1.0	1.0
Below 40	100	0.5	1.0	1.0	1.0	1.5

INSULATION FOR HOT EQUIPMENT

❑ Check specification to determine if insulation is required to be rigid block or semi-rigid board.
❑ Check for specified type of material and thickness of insulation being installed.
❑ Form or fabricate insulation to fit equipment.
❑ On round equipment, insulation edges will be beveled to ensure tight joints.

- ❏ Check joints for being tightly butted, being filled with mineral fiber, or insulation cement.
- ❏ Check specifications and manufacturers' recommendation on spacing of bands. Spacing will not be less than 12 in. on centers.
- ❏ Check for excessive use of wires in lieu of bands. Check for insulation protectors under wires.
- ❏ Check hot ducts and equipment for specified finish.
- ❏ Check for continuity of insulation through walls and floors.

INSULATION FOR COLD EQUIPMENT

- ❏ Check dual temperature equipment, which operates at 60°F or below at any time, for insulation as specified for cold equipment. Check specification for pump insulation. It may vary from flexible, rigid, or semirigid type insulation. Check all other equipment for the specified insulation.
- ❏ Check insulation for the thickness specified.
- ❏ Check installation of vapor barrier.
- ❏ Check drain pans under pumps for insulation underneath.
- ❏ Check cold duct and equipment insulation finish, in accordance with specifications.

ABOVEGROUND PIPE INSULATION

❏ Check contract specifications to determine type of insulation required on pipelines within the structure.
  (a) Normally, domestic hot water, steam, condensate, hot water heating, heated oil, and water defrost lines are insulated as hot pipelines.
  (b) Normally domestic cold water, interior roof drains, refrigerant suction lines, chilled water and dual temperature water lines, air-conditioner condensate drain pipelines exposed to weather drainage piping, and piping which operates at 60°F or below at any time are insulated as cold pipelines.
❏ Check exterior piping for insulation as required by specifications for piping exposed to weather.
❏ Check specifications for areas which are to receive factory-applied vapor barrier jackets, field-applied aluminum jackets, and field-applied vapor barriers.

PIPING EXPOSED TO WEATHER

❏ Check to see that the pipe is insulated and jacketed for applicable service. Note that the vapor barrier is not normally specified for hot pipelines.

❑ Check to see if specified jacket is aluminum.

❑ Check to see if the jacket is required to be factory applied or field applied.

   (a) Check to see if the aluminum jacket laps not less than 2 in. at all joints.

   (b) Check banding requirements for the jacket.

   (c) Check to see that horizontal joints are lapped downward to shed water, and that vertical joints are sealed with a waterproof coating.

❑ Check specifications for special treatment of flanges, couplings, unions, valves, fittings, and anchors.

UNDERGROUND PIPE INSULATION

❑ Check all below-ground domestic hot water heating, heating hot water to 200°F, dual temperature water, and chilled water piping for specified insulation. Generally, the insulation is 1½-in.-thick cellular glass.

Cellular Glass Insulation

❑ Check to see that bore surfaces of insulation are coated with a thin application of high-strength gypsum cement, as recommended by the manufacturer.

❑ Insulation Joint Checklist
- Staggered, one-half overlapping the next opposite half section.
- Tightly butted and seated with bedding compound.
- Insulation secured with two stainless steel bands per section of insulation.
- Insulation terminates at anchor blocks.
- Insulation is continuous through sleeves and manhole.
- Backfill around and 3 in. above the insulation to be free of stones larger than $\frac{1}{4}$-in. any dimension.
- Insulation extends 2 in. inside building interior and tightly butted, scaled, and vapor barrier coated to interior piping.
- Check for special insulation requirements for flanges, couplings, unions, valves, and fittings.
- Check finish of insulation for two coats of mastic with glass cloth or tape embedded between coats. Check for proper overlap at all joints.

❑ Check wet film thickness of both coats of mastic to meet specifications requirements.

❑ Check termination points to see that mastic and cloth or tape covers the end of the insulation and extends along the base pipe as required by the specifications.

CHAPTER 6
HEATING SYSTEMS

GENERAL

This chapter covers material, equipment, and good workmanship practices for the installation of heating systems.

MATERIALS AND EQUIPMENT

❑ Make sure that each piece of material and each item of equipment have been approved well in advance of their need. When the material and equipment arrive on the job, inspect them very carefully, comparing them with the approved shop drawing and samples. Check and record nameplate data on all equipment.

❑ Determine that there is adequate space in the room for proper functioning and maintenance of all the equipment.

❑ Reject all damaged materials and equipment and have them removed from the site.

❑ Check the electrical features of equipment and coordinate with the mechanical features.

❑ Determine that provisions have been made for access panels.

- ❑ Check specification provisions for necessary spare parts and tools for all of the equipment.
- ❑ Require proper storage and protection of all materials and equipment.
- ❑ Check the required controls and valves for compliance with contract requirements.
- ❑ Check the noise level of all equipment.
- ❑ Verify requirements for the installation of flexible pipe connections and vibration eliminators for equipment.
- ❑ Check the installation of all equipment for compliance with manufacturer's recommendations.
- ❑ See that operations and maintenance instructions are with respective equipment and are posted on the wall upon completion of installation.

Boilers, Furnaces, and Accessory Equipment

- ❑ Examine pressure boilers for conformance with the ASME Code.
- ❑ Check for all necessary connections on the boiler.
- ❑ Check cast iron boilers, if field assembled, for tightness of joints. All joints must be sealed.
- ❑ Reject cracked sections.

- ❏ Inspect refractory furnaces built up on the job for materials and workmanship.
- ❏ Require expansion joints. Piping on both sides of expansion joints must be properly braced.
- ❏ Ensure packing to prevent gas or air leakage.
- ❏ Reject all cracked, chipped, or otherwise damaged brick and tile.
- ❏ Check plastic refractories for placement, thorough ramming, and consistency.
- ❏ Require that refractories be kept dry.
- ❏ Inspect for use of refractory mortar in construction of combustion chamber.
- ❏ Check for air circulation under the combustion chamber floors.
- ❏ Inspect the application of insulation after all joints are tightly sealed. Check material, thickness, and finish.
- ❏ Observe accessory equipment operation such as feedwater controllers, dampers, pressure and draft gages, flow and pressure recorders, soot blowers, water columns, and boiler blowdown.
- ❏ Check the pressurestat differential.
- ❏ Check requirement for expansion joint in floor around boiler.
- ❏ Before rolling in, check the cleaning of ends of tubes and the surfaces of tube holes in drums and headers. Check to ensure that

new boilers exposed to weather are covered to prevent corrosion.

❏ See that tube-rolling is done by experienced workmen and that all precautions are taken to prevent either under- or over-rolling. At this stage of erection, request technical assistance.

❏ Ensure that the boiler inspector is notified when it is time for the hydrostatic test. Obtain a Certificate of Inspection. Do not permit the installation of any baffles or the setting of refractories until after the boiler has passed inspection.

❏ Affirm that baffles of steel, refractory tile, or monolithic construction are installed gas-tight but with provision for expansion, and that they will resist dislodgment by "puffs."

❏ Ensure that boiling-out operations for the removal of grease, oil, and other foreign matter are performed before the boiler is placed in operation.

❏ Ensure that space is provided for tube removal and cleaning and for general maintenance of all equipment.

❏ Check to ensure that during periods of operation by the contractor, chemical treatment and blowdown are provided to prevent scale deposits and corrosion.

- ❏ Be sure that all settings are constructed with provision for expansion and contraction of both the refractories and the pressure parts. See that expansion joints are sealed to prevent passage of air or gases but are flexible enough to maintain their seal under movement of the structure. Check the entire setting for leaks.
- ❏ Check solid refractory walls for plumb, level courses, and dipped joints. Check grades of refractories used. Chipped, cracked, wet, or broken refractory materials will be rejected.
- ❏ Ensure that refractory tile and setting casings are constructed to prevent the escape of gases or the infiltration of air, and that they are installed in accordance with the recommendations of the manufacturer.
- ❏ Insist that all openings through setting walls are accurately located and of proper size. Check temperature of the boiler setting surface against room temperature. Verify that pipe sleeves for draft gages are clean and flush with the interior face of the wall.
- ❏ Inspect uptake damper for correct location, bearing material, and freedom of operation when hot.

Draft Fans and Ductwork

❑ Check that induced draft fans are provided with cleanout doors.
❑ Verify the operation of dampers at high flue gas temperatures.

TABLE 6-1	BOILER/WATER HEATERS EXPANSION TANK	
System Volume	Diaphragm Type	Nonpressurized Type
10	1.0	1.5
20	1.5	3.0
30	2.5	4.5
40	3.0	6.0
50	4.0	7.5
60	5.0	9.0
70	6.0	10.5
80	6.5	12.0
90	7.5	13.5
100	8.0	15.0

Note: Minimum Capacities (Water temperature 195°F, fill pressure 12 psig, and minimum operating pressure 30 psig.)

Fuel Burning Equipment

Oil Burners Checklist

- ❏ Size and type of burner tips
- ❏ Location of electrodes to ensure spark in oil spray cone
- ❏ Position of gas or oil pilot
- ❏ Clearances for removal of burner from furnace
- ❏ Burner adjustments
- ❏ Carbon dioxide in flue gas

Gas Burners

- ❏ Check for cleanliness, adjustments, position of pilot flame, and sensing element. Check regulator and controls.
- ❏ Blow out gas line before connecting to burner or regulator.
- ❏ Install regulator in vertical position.
- ❏ Pipe the gas vents to the outdoors.

FUEL-BURNING APPLIANCES

Fuel-burning appliances must have the following characteristics and dimensions.

- Must be accessible without moving permanent structure.
- Must have 30-in platform (working space) in

front of control side of appliance. (**Exception:** room heaters need only 18 in. of working space.)
- Not less than 1 in. clear air space from combustible materials.
- ***All must have a label!***

TABLE 6-2	CLEARANCES FOR CENTRAL FURNACES			
Type	Top	Front	Back	Sides
Oil	24 in.	24 in.	12 in.	36 in.
Gas	18 in.	18 in.	12 in.	36 in.
Elect.	18 in.	18 in.	24 in.	36 in.

APPLIANCE TYPES

Category I These appliances operate with a nonpositive vent connection pressure and with a flue gas temperature of at least 140°F above its dewpoint.

Category II These appliances operate with a nonpositive vent connection pressure and with a flue gas temperature less than 140°F above its dewpoint.

Yellow-tipped flames
- Yellow-tipped flames with a small inner bluish flame at the burner ports usually indicates incomplete combustion of the gas due to insufficient supply of primary air. Soot will usually form on vents and heat exchangers.

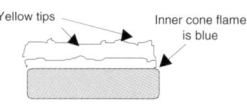

Yellow tips   Inner cone flame is blue

Yellow-tipped flame

Lazy-soft flame
- This flame occurs when the primary is sufficient enough to eliminate the yellow tips on the outer portion of the flame. Both the inner and outer flames are not well defined.

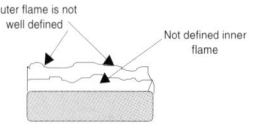

Outer flame is not well defined   Not defined inner flame

Lazy-soft flame

Floating flame
- Indicates insufficient secondary air supply which is caused by a restricted air flow, bad venting, or a combination of the two. This flame usually gives an odor because of the unburned gas by-products.

Figure 6-1   FLAME TYPES

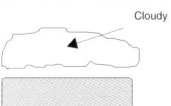

Cloudy

Floating flame

Uniform lifting flame
- Caused by excessive amounts of primary air being supplied to the burner usually can also be heard as a blowing noise.

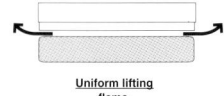

Uniform lifting
flame

Orange-colored flame
- Usually caused by dust burning in the flame— usually not a problem unless it's excessive.

Red streaks

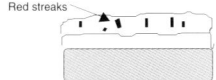

Orange-like flame

Sharply defined flame
- Good combustion. A sharp, crisp flame indicates proper mixture of gas, secondary and primary airs.
- Both inner and outer have straight sides, with the inner flame resting on the burner ports with no clearance between them.

Category III These appliances operate with **a positive** vent pressure and with a flue gas temperature of **at least 140°F** above its dewpoint.

Category IV These appliances operate with **a positive** vent pressure and with a flue gas temperature **less than 140°F** above its dewpoint.

Appliance Installations

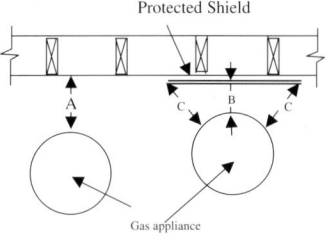

- **A** is equal to clearance allowed with no protection.
- **B** is the "reduced clearance" as shown in Tables 6-3 to 6-7.
- **C** is the distance the protective method must extend and equals the original unprotected distance A.

Figure 6-2 CLEARANCES FOR FUEL-BURNING APPLIANCES

TABLE 6-3 TYPES OF CODE-APPROVED APPLIANCE PROTECTION

0	3.5-in.-thick masonry without air space
1	0.5-in.-thick insulation board over 1-in. glass fiber or mineral wool
2	0.024-thick sheet metal over 1-in. glass fiber or mineral wool reinforced with wire on the rear face and with a ventilated air space. (Spacers are to be noncombustionable and not located directly behind the appliance or vent
3	3.5-in. masonry with air space
4	0.024-in.-thick sheet metal with ventilating air space. (At least 1 in. ventilating air space.)
5	0.5-in. thick insulation board with at least 1 in. air space.
6	0.024 sheet metal with 1 in. air space over additional 0.024 piece of sheet metal.
7	1-in. glass fiber or mineral wool sandwiched between two sheets of 0.024-in.-thick sheet metal with ventilating air space

Note: Mineral wool batts if used, are to have a density of 8 lb/ft^3 and a minimum melting point of 1500°F (816 C).

TABLE 6-4 A		
Type of Protection	Above (in.)	Sides and Rear (in)
0	N/A	24
1	24	18
2	18	12
3	N/A	12
4	18	12
5	18	12
6	18	12
7	18	12

Note: Original required clearance Is 36 in.

TABLE 6-5 A		
Type of Protection	Above (in)	Sides and Rear (in)
0	N/A	12
1	12	9
2	9	6
3	N/A	6
4	9	6
5	9	6
6	9	6
7	9	6

Note: Original required clearance is 18 in.

TABLE 6-6	A	
Type of Protection	Above (in.)	Sides and Rear (in)
0	N/A	9
1	9	6
2	6	4
3	N/A	6
4	6	4
5	6	4
6	6	4
7	6	4

Note: Original required clearance is 12 in.

TABLE 6-7	A	
Type of Protection	Above (in.)	Sides and Rear (in)
0	N/A	6
1	6	5
2	5	3
3	N/A	6
4	5	3
5	5	3
6	5	3
7	5	3

Note: Original required clearance is 9 in.

HEAT

Type of Protection	Above (in.)	Sides and Rear (in)
0	N/A	5
1	4	3
2	3	3
3	N/A	6
4	3	2
5	3	3
6	3	3
7	3	3

TABLE 6-8 A

Note: Original required clearance is **6** in.

Clothes Dryer Duct

Ducts must be galvanized steel or aluminum if passing through any fire-rated assembly.

- Ducts must terminate outside and not be connected to any other vent assembly.
- Ducts cannot exceed 25 ft in length.
  Reduce:
  (a) 2½ ft for each 45° bend
  (b) 5 ft for each 90° bend
- The duct must be a minimum of 4 in. in diameter.

See "Up to Code" Plumbing for more information on clothes dryers.

- Joints must be in the direction of the flow.
- Exhaust vents need a **backdraft damper.**
- Ducts to be of rigid metal with no screws penetrating to the interior of the duct.
- Length cannot **exceed 25** ft.
- Reduce overall length by **2.5** ft for every 45° bend and **5′** for each 90° bend.
- Gas dryers **cannot** be located in a room with other fuel burning appliances.

Decorative Appliances

A decorative gas appliance is a vented appliance installed for the aesthetic effect of the flames rather than functional effects.

TABLE 6-9 DAMPER OPEN AREA FROM UNLISTED DECORATIVE APPLIANCES

Minimum Permanent Free Opening	Chimney Height (ft)		
	6	8	10
8	7800	8400	9000
13	14000	15200	16800
20	23200	25200	27600
29	34000	37000	40400
39	46400	50400	55800
51	62400	68000	74400
64	80000	86000	96400
	Chimney Height (ft)		
	15	20	30
8	9800	10600	11200
13	18200	20200	21600
20	30200	32600	36600
29	44600	50400	55200
39	62400	68400	76800
51	84000	94000	105800
64	108800	122200	138600

VENTING

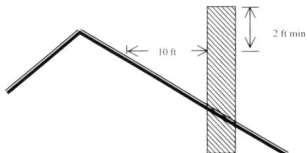

Figure 6-3 ROOF TERMINATIONS FOR CHIMNEYS AND SINGLE WALL VENTS (NO CAPS)

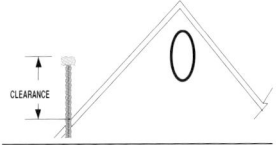

Roof slope	Clearance
Flat–6/12	1′–0 in.
6/12–7/12	1′–3 in
7/12–8/12	1′–6 in.
8/12–9/12	2′–0 in.
9/12–10/12	2′–6 in.
10/12–11/12	3′–3 in.
11/12–12/12	4′–0 in.

Figure 6-4 GAS VENT TERMINATIONS FOR LISTED VENT CAPS (12 IN. OR LESS IN DIAMETER AND 8 FT FROM VERTICAL WALLS)

HEAT

TABLE 6-10 CONVERSIONS FOR FLUE/VENT LINERS

Nominal Liner Size	Inside Liner Dimensions	Circular Equivalent (in.)	Square Inches
		4	12.2
4 × 8	2½ × 6½	5	19.6
		6	28.3
		7	38.3
8 × 8	6¾ × 6¾	7.4	42.7
		8	50.3
8 × 12	6½ × 10½	9	63.6
		10	78.5
12 × 12	9¾ × 9¾	10.4	83.3
		11	95
12 × 16	9½ × 13½	11.8	107.5
		12	113.0
		14	153.9
16 × 16	13¼ × 13¼	14.5	162.9
		15	176.7
16 × 20	13 × 17	16.2	206.1
		18	254.4
20 × 20	16¾ × 16¾	18.2	260.2
		20	314.1
20 × 24	16½ × 20½	20.1	314.2
		22	380.1
24 × 24	20¼ × 20¼	22.1	380.1
		24	452.3

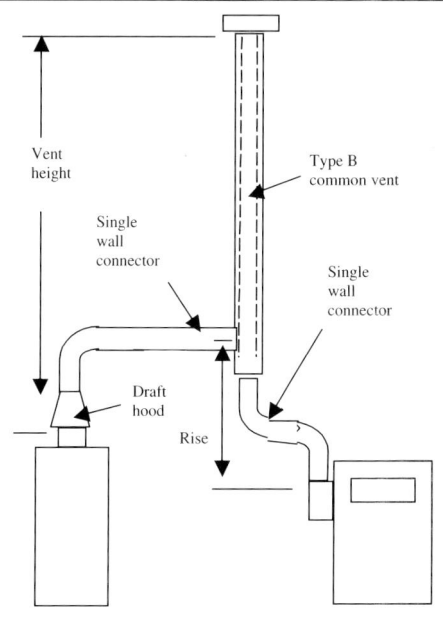

Figure 6-5 MULTIPLE SINGLE WALL CONNECTORS VENT TO TYPE B VENT

Vent height

Type B common vent

Single wall connector

Single wall connector

Draft hood

Rise

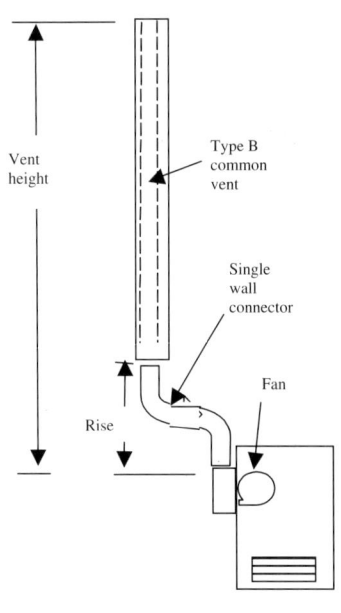

Figure 6-6   FAN ASSISTED VENT

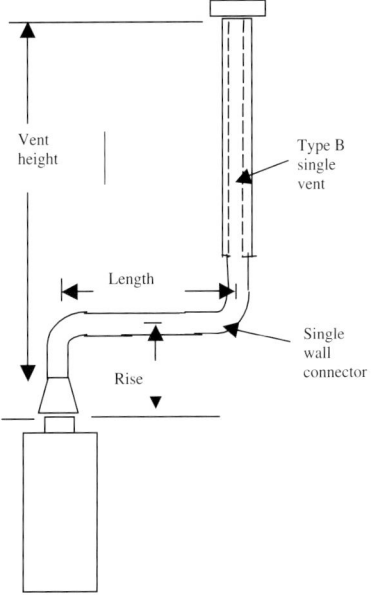

Figure 6-7   SINGLE WALL CONNECTOR VENT TO TYPE B VENT

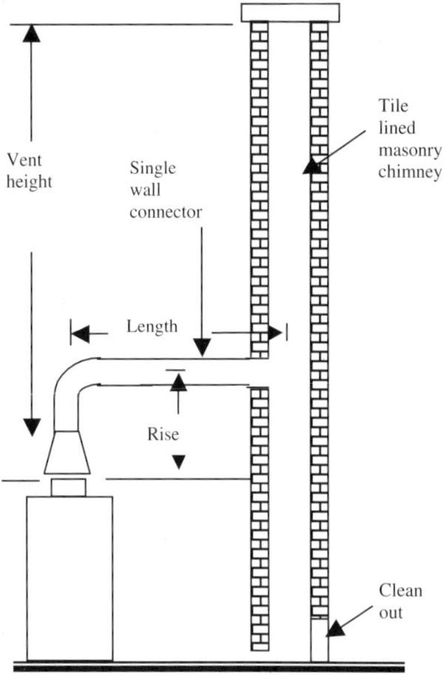

Figure 6-8   **SINGLE WALL VENT TO MASONRY CHIMNEY**

VENT AND CONNECTOR TABLES*

*The following tables are to be used along with the previous section figures.

TABLE 6-11 CAPACITY OF TYPE B DOUBLE-WALL GAS VENTS WHEN CONNECTED DIRECTLY TO A SINGLE CATEGORY I APPLIANCE

Height H (ft)	Lateral L (ft)	Vent Diameter—D (in.)								
		3			4			5		
		Appliance Input Rating in Thousands of Btu/h								
		FAN		NAT	FAN		NAT	FAN		NAT
		Min	Max	Max	Min	Max	Max	Min	Max	Max
6	0	0	78	46	0	152	86	0	251	141
	2	13	51	36	18	97	67	27	157	105
	4	21	49	34	30	94	64	39	153	103
	6	25	46	32	36	91	61	47	149	100
8	0	0	84	50	0	165	94	0	276	155
	2	12	57	40	16	109	75	25	178	120
	5	23	53	38	32	103	71	42	171	115
	8	28	49	35	39	98	66	51	164	109
10	0	0	88	53	0	175	100	0	295	166
	2	12	61	42	17	118	81	23	194	129
	5	23	57	40	32	113	77	41	187	124
	10	30	51	36	41	104	70	54	176	115
15	0	0	94	58	0	191	112	0	327	187
	2	11	69	48	15	136	93	20	226	150
	5	22	65	45	30	130	87	39	219	142
	10	29	59	41	40	121	82	51	206	135
	15	35	53	37	48	112	76	61	195	128

(continued)

HEAT

TABLE 6-11 (CONTINUED)

Height H (ft)	Lateral L (ft)	Vent Diameter—D (in.)							
		3		4			5		
		Appliance Input Rating in Thousands of Btu/h							
		FAN	NAT	FAN		NAT	FAN		NAT
		Min	Max	Max	Min	Max	Max	Min	Max	Max
20	0	0	97	61	0	202	119	0	349	202
	2	10	75	51	14	149	100	18	250	166
	5	21	71	48	29	143	96	38	242	160
	10	28	64	44	38	133	89	50	229	150
	15	34	58	40	46	124	84	59	217	142
	20	48	52	35	55	116	78	69	206	134
30	0	0	100	64	0	213	128	0	374	220
	2	9	81	56	13	166	112	14	283	185
	5	21	77	54	28	160	108	36	275	176
	10	27	70	50	37	150	102	48	262	171
	15	33	64	NA	44	141	96	57	249	163
	20	56	58	NA	53	132	90	66	237	154
	30	NA	NA	NA	73	113	NA	88	214	NA
50	0	0	101	67	0	216	134	0	397′	232
	2	8	86	61	11	183	122	14	320	206
	5	20	82	NA	27	177	119	35	312	200
	10	26	76	NA	35	168	114	45	299	190
	15	59	70	NA	42	158	NA	54	287	180
	20	NA	NA	NA	50	149	NA	63	275	169
	30	NA	NA	NA	69	131	NA	84	250	NA

Height H (ft)	Lateral L (ft)	Vent Diameter—D (in.)								
		6			7			8		
		Appliance Input Rating in Thousands of Btu/h								
		FAN		NAT	FAN		NAT	FAN		NAT
		Min	Max	Max	Min	Max	Max	Min	Max	Max
6	0	0	375	205	0	524	285	0	698	370
	2	32	232	157	44	321	217	53	425	285
	4	50	227	153	66	316	211	79	419	279
	6	59	223	149	78	310	205	93	413	273
8	0	0	415	235	0	583	320	0	780	415
	2	28	263	180	42	365	247	0	483	322
	5	53	255	173	70	356	237	83	473	313
	8	64	247	165	84	347	227	99	463	303
10	0	0	447	255	0	631	345	0	847	450
	2	26	289	195	40	402	273	48	533	355
	5	52	280	188	68	392	263	81	522	346
	10	67	267	175	88	376	245	104	504	330
15	0	0	502	285	0	716	390	0	970	525
	2	22	339	225	38	475	316	45	633	414
	5	49	330	217	64	463	300	76	620	403
	10	64	315	208	84	445	288	99	600	386
	15	76	301	198	98	429	275	115	580	373
20	0	0	540	307	0	776	430	0	1,057	575
	2	20	377	249	33	531	346	41	711	470
	5	47	367	241	62	519	337	73	697	460
	10	62	351	228	81	499	321	95	675	443
	15	73	337	217	94	481	308	111	654	427
	20	84	322	206	107	464	295	125	634	410
30	0	0	587	336	0	853	475	0	1,173	650
	2	18	432	280	27	613	394	33	826	535
	5	45	421	273	58	600	385	69	811	524
	10	59	405	261	77	580	371	91	788	507
	15	70	389	249	90	560	357	105	765	490
	20	80	374	237	102	542	343	119	743	473
	30	104	346	219	131	507	321	149	702	444

(continued)

TABLE 6-11 (CONTINUED)

Height H (ft)	Lateral L (ft)	Vent Diameter—D (in.)								
		6			7			8		
		Appliance Input Rating in Thousands of Btu/h								
		FAN		NAT	FAN		NAT	FAN		NAT
		Min	Max	Max	Min	Max	Max	Min	Max	Max
50	0	0	633	363	0	932	518	0	1,297	708
	2	15	497	314	22	715	445	26	975	615
	5	43	487	308	55	702	438	65	960	605
	10	56	471	298	73	681	426	86	935	589
	15	66	455	288	85	662	413	100	911	572
	20	76	440	278	97	642	401	113	888	556
	30	99	410	259	123	605	376	141	844	522

TABLE 6-12 CAPACITY OF TYPE B DOUBLE-WALL VENTS WITH SINGLE-WALL METAL CONNECTORS SERVING A SINGLE CATEGORY I APPLIANCE

		Vent Diameter—D (in.)											
		3			4			5			6		
		Appliance Input Rating in Thousands of Btu/h											
Height H (ft)	Lateral L (ft)	FAN		NAT	FAN		NAT	FAN		NAT	FAN		NAT
		Min	Max	Max	Min	Max	Max	Min	Max	Max	Min	Max	Max
	0	38	77	45	59	151	85	85	249	140	126	373	204
	2	39	51	36	60	96	66	85	156	104	123	231	156
	4	NA	NA	33	74	92	63	102	152	102	146	225	152
	6	NA	NA	31	83	89	60	114	147	99	163	220	148
	0	37	83	50	58	164	93	83	273	154	123	412	234
	2	39	56	39	59	108	75	83	176	119	121	261	179
	5	NA	NA	37	77	102	69	107	168	114	151	252	171
	8	NA	NA	33	90	95	64	122	161	107	175	243	163
	0	37	87	53	57	174	99	82	293	165	120	444	254
	2	39	61	41	59	117	80	82	193	128	119	287	194
	5	52	56	39	76	111	76	105	185	122	148	277	186
	10	NA	NA	34	97	100	68	132	171	112	188	261	171
	0	36	93	57	56	190	111	80	325	186	116	499	283
	2	38	69	47	57	136	93	80	225	149	115	337	224
15	5	51	63	44	75	128	86	102	216	140	144	326	217
	10	NA	NA	39	95	116	79	128	201	131	182	308	203
	15	NA	NA	NA	NA	NA	72	158	186	124	220	290	192
	0	35	96	60	54	200	118	78	346	201	114	537	306
	2	37	74	50	56	148	99	78	248	165	113	375	248
	5	50	68	47	73	140	94	100	239	158	141	363	239
	10	NA	NA	41	93	129	86	125	223	146	177	344	224
	15	NA	NA	NA	NA	NA	80	155	208	136	216	325	210
	20	NA	NA	NA	NA	NA	NA	186	192	126	254	306	196

(continued)

TABLE 6-12 (CONTINUED)

Height H (ft)	Lateral L (ft)	Vent Diameter—D (in.)										
		3		4			5			6		
		Appliance Input Rating in Thousands of Btu/h										
		FAN		NAT	FAN		NAT	FAN		NAT	FAN		NAT
		Min	Max	Max	Min	Max	Max	Min	Max	Max	Min	Max	Max
30	0	34	99	63	53	211	127	76	372	219	110	584	334
	2	37	80	56	55	164	111	76	281	183	109	429	279
	5	49	74	52	72	157	106	98	271	173	136	417	271
	10	NA	NA	NA	91	144	98	122	255	168	171	397	257
	15	NA	NA	NA	115	131	NA	151	239	157	208	377	242
	20	NA	NA	NA	NA	NA	NA	181	223	NA	246	357	228
	30	NA	NA	NA	NA	NA	NA	NA	NA	NA	NA	NA	NA
50	0	33	99	66	51	213	133	73	394	230	105	629	361
	2	36	84	61	53	181	121	73	318	205	104	495	312
	5	48	80	NA	70	174	117	94	308	198	131	482	305
	10	NA	NA	NA	89	160	NA	118	292	186	162	461	292
	15	NA	NA	NA	112	148	NA	145	275	174	199	441	280
	20	NA	NA	NA	NA	NA	NA	176	257	NA	236	420	267
	30	NA	NA	NA	NA	NA	NA	NA	NA	NA	315	376	NA

Height H (ft)	Lateral L (ft)	Vent Diameter—D (in.)										
		7			8			9			10		
		Appliance Input Rating in Thousands of Btu/h											
		FAN		NAT	FAN		NAT	FAN		NAT	FAN		NAT
		Min	Max	Max	Min	Max	Max	Min	Max	Max	Min	Max	Max
	0	165	522	284	211	695	369	267	894	469	371	1,118	569
	2	159	320	213	201	423	284	251	541	368	347	673	453
	4	187	313	208	237	416	277	295	533	360	409	664	443
	6	207	307	203	263	409	271	327	526	352	449	656	433
	0	161	580	319	206	777	414	258	1,002	536	360	1,257	658
	2	155	363	246	197	482	321	246	617	417	339	768	513
	5	193	352	235	245	470	311	305	604	404	418'	754	500
	8	223	342	225	280	458	300	344	591	392	470	740	486
	0	158	628	344	202	844	449	253	1,093	584	351	1,373	718
	2	153	400	272	193	531	354	242	681	456	332	849	559
	5	190	388	261	241	518	344	299	667	443	409	834	544
	10	237	369	241	296	497	325	363	643	423	492	808	520
15	0	153	713	388	195	966	523	244	1,259	681	336	1,591	838
	2	148	473	314	187	631	413	232	812	543	319	1,015	673
	5	182	459	298	231	616	400	287	795	526	392	997	657
	10	228	438	284	284	592	381	349	768	501	470	966	628
	15	272	418	269	334	568	367	404	742	484	540	937	601
	0	149	772	428	190	1,053	573	238	1,379	750	326	1,751	927
	2	144	528	344	182	708	468	227	914	611	309	1,146	754
	5	178	514	334	224	692	457	279	896	596	381	1,126	734
	10	222	491	316	277	666	437	339	866	570	457	1,092	702
	15	264	469	301	325	640	419	393	838	549	526	1,060	677
	20	309	448	285	374	616	400	448	810	526	592	1,028	651

(continued)

TABLE 6-12 (CONTINUED)

Height H (ft)	Lateral L (ft)	Vent Diameter—D (in.)											
		7			8			9			10		
		Appliance Input Rating in Thousands of Btu/h											
		FAN		NAT	FAN		NAT	FAN		NAT	FAN		NAT
		Min	Max	Max	Min	Max	Max	Min	Max	Max	Min	Max	Max
30	0	144	849	472	184	1,168	647	229	1,542	852	312	1,971	1,056
	2	139	610	392	175	823	533	219	1,069	698	296	1,346	863
	5	171	595	382	215	806	521	269	1,049	684	366	1,324	846
	10	213	570	367	265	777	501	327	1,017	662	440	1,287	821
	15	255	547	349	312	750	481	379	985	638	507	1,251	794
	20	298	524	333	360	723	461	433	955	615	570	1,216	768
	30	389	477	305	461	670	426	541	895	574	704	1,147	720
50	0	138	928	515	176	1,292	704	220	1,724	948	295	2,223	1,189
	2	133	712	443	168	971	613	209	1,273	811	280	1,615	1,007
	5	164	696	435	204	953	602	257	1,252	795	347	1,591	991
	10	203	671	420	253	923	583	313	1,217	765	418	1,551	963
	15	244	646	405	299	894	562	363	1,183	736	481	1,512	934
	20	285	622	389	345	866	543	415	1,150	708	544	1,473	906
	30	373	573	NA	442	809	502	521	1,086	649	674	1,399	848

TABLE 6-13 VENT CONNECTOR CAPACITY OF TYPE B DOUBLE-WALL VENTS WITH TYPE B DOUBLE-WALL CONNECTORS SERVING TWO OR MORE CATEGORY I APPLIANCES

Height H (ft)	Lateral L (ft)	3				4				6			
		FAN		NAT		FAN		NAT		FAN		NAT	
		Min	Max	Max	Min	Max	Max	Min	Max	Max	Min	Max	Max
6	1	22	37	26	35	66	46	46	106	72	58	164	104
	2	23	41	31	37	75	55	48	121	86	60	183	124
	3	24	44	35	38	81	62	49	132	96	62	199	139
8	1	22	40	27	35	72	48	49	114	76	64	176	109
	2	23	44	32	36	80	57	51	128	90	66	195	129
	3	24	47	36	37	87	64	53	139	101	67	210	145
10	1	22	43	28	34	78	50	49	123	78	65	189	113
	2	23	47	33	36	86	59	51	136	93	67	206	134
	3	24	50	37	37	92	67	52	146	104	69	220	150
15	1	21	50	30	33	89	53	47	142	83	64	220	120
	2	22	53	35	35	96	63	49	153	99	66	235	142
	3	24	55	40	36	102	71	51	163	111	68	248	160
20	1	21	54	31	33	99	56	46	157	87	62	246	125
	2	22	57	37	34	105	66	48	167	104	64	259	149
	3	23	60	42	35	110	74	50	176	116	66	271	168
30	1	20	62	33	31	113	59	45	181	93	60	288	134
	2	21	64	39	33	118	70	47	190	110	62	299	158
	3	22	66	44	34	23	79	48	198	124	64	309	178
50	1	19	71	36	30	133	64	43	216	101	57	349	145
	2	21	73	43	32	137	76	45	223	119	59	358	172
	3	22	75	48	33	141	86	46	229	134	61	366	194

(continued)

TABLE 6-13 (CONTINUED)

Height H (ft)	Lateral L (ft)	Type B Double-Wall Vent and Connector Diameter —D (in.)											
		7			8			9			10		
		Appliance Input Rating in Thousands of Btu/h											
		FAN		NAT	FAN		NAT	FAN		NAT	FAN		NAT
		Min	Max	Max	Min	Max	Max	Min	Max	Max	Min	Max	Max
6	1	77	225	142	92	296	185	109	376	237	128	466	289
	2	79	253	168	95	333	220	112	424	282	131	526	345
	3	82	275	189	97	363	248	114	463	317	134	575	386
8	1	84	243	148	100	320	194	118	408	248	138	507	303
	2	86	269	175	103	356	230	121	454	294	141	564	358
	3	88	290	198	105	384	258	123	492	330	143	612	402
10	1	89	257	154	106	341	200	125	436	257	146	542	314
	2	91	282	182	109	374	238	128	479	305	149	596	372
	3	94	303	205	111	402	268	131	515	342	152	642	417
15	1	88	298	163	110	389	214	134	493	273	162	609	333
	2	91	320	193	112	419	253	137	532	323	165	658	394
	3	93	339	218	115	445	286	140	565	365	167	700	444
20	1	86	334	171	107	436	224	131	552	285	158	681	347
	2	89	354	202	110	463	265	134	587	339	161	725	414
	3	91	371	228	113	486	300	137	618	383	164	764	466
30	1	83	391	182	103	512	238	125	649	303	151	802	372
	2	85	408	215	105	535	282	129	679	360	155	840	439
	3	88	423	242	108	555	317	132	706	405	158	874	494
50	1	78	477	197	97	627	257	120	797	330	144	984	403
	2	81	490	234	100	645	306	123	820	392	148	1,014	478
	3	83	502	263	103	661	343	126	842	441	151	1,043	538

	Type B Double-Wall Vent and Connector Diameter—D (in.)								
	4			**5**			**6**		
Vent Height H (ft)	Combined Appliance Input Rating in Thousands of Btu/h								
	FAN +FAN	FAN +FAN	NAT +NAT	FAN +FAN	FAN +NAT	NAT +NAT	FAN +FAN	FAN +NAT	NAT +NAT
6	92	81	65	140	116	103	204	161	147
8	101	90	73	155	129	114	224	178	163
10	110	97	79	169	141	124	243	194	178
15	125	112	91	195	164	144	283	228	206
20	136	123	102	215	183	160	314	255	229
30	152	138	118	244	210	185	361	297	266
50	167	153	134	279	244	214	421	353	310

	Type B Double-Wall Vent and Connector Diameter—D (in.)								
	7			**8**			**9**		
Vent Height H (ft)	Combined Appliance Input Rating in Thousands of Btu/h								
	FAN +FAN	FAN +NAT	NAT +NAT	FAN +FAN	FAN +NAT	NAT +NAT	FAN +FAN	FAN +NAT	NAT +NAT
6	309	248	200	404	314	260	547	434	335
8	339	275	223	444	348	290	602	480	378
10	367	299	242	477	377	320	649	522	405
15	427	352	280	556	444	365	753	612	465
20	475	394	310	621	499	405	842	688	523
30	547	459	360	720	585	470	979	808	605
50	641	547	423	854	706	550	1.164	977	705

TABLE 6-14   VENT CONNECTOR CAPACITY OF TYPE B DOUBLE-WALL VENT WITH SINGLE-WALL CONNECTORS SERVING TWO OR MORE CATEGORY I APPLIANCES

Vent Height H (ft)	Connector Rise R (ft)	Single-Wall Metal Vent Connector Diameter—D (in)											
		3			4			5			6		
		Appliance Input Rating in Thousands of Btu/h											
		FAN		NAT	FAN		NAT	FAN		NAT	FAN		NAT
		Min	Max	Max	Min	Max	Max	Min	Max	Max	Min	Max	Max
6	1	NA	NA	26	NA	NA	46	NA	NA	71	NA	NA	102
	2	NA	NA	31	NA	NA	55	NA	NA	85	168	182	123
	3	NA	NA	34	NA	NA	62	121	131	95	175	198	138
8	1	NA	NA	27	NA	NA	48	NA	NA	75	NA	NA	106
	2	NA	NA	32	NA	NA	57	125	126	89	184	193	127
	3	NA	NA	35	NA	NA	64	130	138	100	191	208	144
10	1	NA	NA	28	NA	NA	50	119	121	77	182	186	110
	2	NA	NA	33	84	85	59	124	134	91	189	203	132
	3	NA	NA	36	89	91	67	129	144	102	197	217	148
15	1	NA	NA	29	79	87	52	116	138	81	177	214	116
	2	NA	NA	34	83	94	62	121	150	97	185	230	138
	3	NA	NA	39	87	100	70	127	160	109	193	243	157
20	1	49	56	30	78	97	54	115	152	84	175	238	120
	2	52	59	36	82	103	64	120	163	101	182	252	144
	3	55	62	40	87	107	72	125	172	113	190	264	164
30	1	47	60	31	77	110	57	112	175	89	169	278	129
	2	51	62	37	81	115	67	117	185	106	177	290	152
	3	54	64	42	85	119	76	122	193	120	185	300	172
50	1	46	69	34	75	128	60	109	207	96	162	336	137
	2	49	71	40	79	132	72	114	215	113	170	345	164
	3	52	72	45	83	136	82	119	221	123	178	353	186

Vent Height H (ft)	Connector Rise R (ft)	Single-Wall Metal Vent Connector Diameter—D (in)											
		7			8			9			10		
		Appliance Input Rating in Thousands of Btu/h											
		FAN		NAT	FAN		NAT	FAN		NAT	FAN		NAT
		Min	Max	Max	Min	Max	Max	Min	Max	Max	Min	Max	Max
6	1	207	223	140	262	293	183	325	373	234	447	463	286
	2	215	251	167	271	331	219	334	422	281	458	524	344
	3	222	273	188	279	361	247	344	462	316	468	574	385
8	1	226	240	145	285	316	191	352	403	244	481	502	299
	2	234	266	173	293	353	228	360	450	292	492	560	355
	3	241	287	197	302	381	256	370	489	328	501	609	400
10	1	240	253	150	302	335	196	372	429	252	506	534	308
	2	248	278	183	311	369	235	381	473	302	517	589	368
	3	257	299	203	320	398	265	391	511	339	528	637	413
15	1	238	291	158	312	380	208	397	482	266	556	596	324
	2	246	314	189	321	411	248	407	522	317	568	646	387
	3	255	333	215	331	438	281	418	557	360	579	690	437
20	1	233	325	165	306	425	217	390	538	276	546	664	336
	2	243	346	197	317	453	259	400	574	331	558	709	403
	3	252	363	223	326	476	294	412	607	375	570	750	457
30	1	226	380	175	296	497	230	378	630	294	528	779	358
	2	236	397	208	307	521	274	389	662	349	541	819	425
	3	244	412	235	316	542	309	400	690	394	555	855	482
50	1	217	460	188	284	604	245	364	768	314	507	951	384
	2	226	473	223	294	623	293	376	793	375	520	983	458
	3	235	486	252	304	640	331	387	816	423	535	1,013	518

HEAT

TABLE 6-15 COMMON VENT CAPACITY OF TYPE B DOUBLE-WALL VENT WITH SINGLE-WALL CONNECTORS SERVING TWO OR MORE CATEGORY I APPLIANCES

Vent Height H (ft)	Type B Double-Wall Vent and Diameter—D (in)								
	4			5			6		
	Combined Appliance Input Rating in Thousands of Btu/h								
	FAN +FAN	FAN +NAT	NAT +NAT	FAN +FAN	FAN +NAT	NAT +NAT	FAN +FAN	FAN +NAT	NAT +NAT
6	NA	78	64	NA	113	99	200	158	144
8	NA	87	71	NA	126	111	218	173	159
10	NA	94	76	163	137	120	237	189	174
15	121	108	88	189	159	140	275	221	200
20	131	118	98	208	177	156	305	247	223
30	145	132	113	236	202	180	350	286	257
50	159	145	128	268	233	208	406	337	296

Vent Height H (ft)	Type B Double-Wall Vent and Diameter—D (in)								
	7			8			9		
	Combined Appliance Input Rating in Thousands of Btu/h								
	FAN +FAN	FAN +NAT	NAT +NAT	FAN +FAN	FAN +NAT	NAT +NAT	FAN +FAN	FAN +NAT	NAT +NAT
6	304	244	196	398	310	257	541	429	332
8	331	269	218	436	342	285	592	473	373
10	357	292	236	467	369	309	638	512	398
15	416	343	274	544	434	357	738	599	456
20	463	383	302	606	487	395	824	673	512
30	533	446	349	703	570	459	958	790	593
50	622	529	410	833	686	535	1,139	954	689

TABLE 6-16 VENT CONNECTOR CAPACITY OF MASONRY CHIMNEY WITH TYPE B DOUBLE-WALL CONNECTORS SERVING TWO OR MORE CATEGORY I APPLIANCES

		Type B Double-Wall Vent Connector Diameter—D (in)										
		3			4			5			6		
Vent Height H (ft)	Connector Rise R (ft)	Appliance Input Rating in Thousands of Btu/h											
		FAN		NAT	FAN		NAT	FAN		NAT	FAN		NAT
		Min	Max	Max	Min	Max	Max	Min	Max	Max	Min	Max	Max
6	1	24	33	21	39	62	40	52	106	67	65	194	101
	2	26	43	28	41	79	52	53	133	85	67	230	124
	3	27	49	34	42	92	61	55	155	97	69	262	143
8	1	24	39	22	39	72	41	55	117	69	71	213	105
	2	26	47	29	40	87	53	57	140	86	73	246	127
	3	27	52	34	42	97	62	59	159	98	75	269	145
10	1	24	42	22	38	80	42	55	130	71	74	232	108
	2	26	50	29	40	93	54	57	153	87	76	261	129
	3	27	55	35	41	105	63	58	170	100	78	284	148
15	1	24	48	23	38	93	44	54	154	74	72	277	114
	2	25	55	31	39	105	55	56	174	89	74	299	134
	3	26	59	35	41	115	64	57	189	102	76	319	153
20	1	24	52	24	37	102	46	53	172	77	71	313	119
	2	25	58	31	39	114	56	55	190	91	73	335	138
	3	26	63	35	40	123	65	57	204	104	75	353	157
30	1	24	54	25	37	111	48	52	192	82	69	357	127
	2	25	60	32	38	122	58	54	208	95	72	376	145
	3	26	64	36	40	131	66	56	221	107	74	392	163
50	1	23	51	25	36	116	51	51	209	89	67	405	143
	2	24	59	32	37	127	61	53	225	102	70	421	161
	3	26	64	36	39	135	69	55	237	115	72	435	180

(continued)

HEAT

TABLE 6-16 (CONTINUED)

Vent Height H (ft)	Connector Rise R (ft)	Type B Double-Wall Vent Connector Diameter—D (in)										
		7			8			9			10		
		Appliance Input Rating in Thousands of Btu/h											
		FAN		NAT	FAN		NAT	FAN		NAT	FAN		NAT
		Min	Max	Max	Min	Max	Max	Min	Max	Max	Min	Max	Max
6	1	87	274	141	104	370	201	124	479	253	145	599	319
	2	89	324	173	107	436	232	127	562	300	148	694	378
	3	91	369	203	109	491	270	129	633	349	151	795	439
8	1	94	304	148	113	414	210	134	539	267	156	682	335
	2	97	350	179	116	473	240	137	615	311	160	776	394
	3	99	383	206	119	517	276	139	672	358	163	848	452
10	1	101	324	153	120	444	216	142	582	277	165	739	348
	2	103	366	184	123	498	247	145	652	321	168	825	407
	3	106	397	209	126	540	281	147	705	366	171	893	463
15	1	100	384	164	125	511	229	153	658	297	184	824	375
	2	103	419	192	128	558	260	156	718	339	187	900	432
	3	105	448	215	131	597	292	159	760	382	190	960	486
20	1	98	437	173	123	584	239	150	752	312	180	943	397
	2	101	467	199	126	625	270	153	805	354	184	1011	452
	3	104	493	222	129	661	301	156	851	396	187	1067	505
30	1	96	504	187	119	680	255	145	883	337	175	1,115	432
	2	99	528	209	122	715	287	149	928	378	179	1,171	484
	3	101	554	233	125	746	317	152	968	418	182	1,220	535
50	1	92	582	213	115	798	294	140	1,049	392	168	1,334	506
	2	95	604	235	118	827	326	143	1,085	433	172	1,379	558
	3	98	624	260	121	854	357	147	1,118	474	176	1,421	611

	Mimimum Internal Area of Masonry Chimney Flue (sq in.)										
	12			19			28			38		
Vent Height H (ft)	Combined Appliance Input Rating in Thousands of Btu/h											
	FAN +FAN	FAN +FAN	NAT +NAT	FAN +FAN	FAN +FAN	NAT +NAT	FAN +FAN	FAN +FAN	NAT +NAT	FAN +FAN	FAN +FAN	NAT +NAT
6	NA	74	25	NA	119	46	NA	178	71	NA	257	103
8	NA	80	28	NA	130	53	NA	193	82	NA	279	119
10	NA	84	31	NA	138	56	NA	207	90	NA	299	131
15	NA	NA	36	NA	152	67	NA	233	106	NA	334	152
20	NA	NA	41	NA	NA	75	NA	250	122	NA	368	172
30	NA	NA	NA	NA	NA	NA	NA	270	137	NA	404	198
50	NA	NA	NA	NA	NA	NA	NA	NA	NA	NA	NA	NA

	Mimimum Internal Area of Masonry Chimney Flue (sq in.)										
	50			63			78			113		
Vent Height H (ft)	Combined Appliance Input Rating in Thousands of Btu/h											
	FAN +FAN	FAN +FAN	NAT +NAT	FAN +FAN	FAN +FAN	NAT +NAT	FAN +FAN	FAN +FAN	NAT +NAT	FAN +FAN	FAN +FAN	NAT +NAT
6	NA	351	143	NA	458	188	NA	582	246	1,041	853	NA
8	NA	384	163	NA	501	218	724	636	278	1,144	937	408
10	NA	409	177	606	538	236	776	686	302	1,226	1,010	454
15	523	467	212	682	611	283	874	781	365	1,374	1,156	546
20	565	508	243	742	668	325	955	858	419	1,513	1,286	648
30	615	564	278	816	747	381	1,062	969	496	1,702	1,473	749
50	NA	620	328	879	831	461	1,165	1,089	606	1,905	1,692	922

TABLE 6-17 VENT CONNECTOR CAPACITY OF MASONRY CHIMNEY WITH SINGLE-WALL CONNECTORS SERVING TWO OR MORE CATEGORY I APPLIANCES

Vent Height H (ft)	Connector Rise R (ft)	3			4			5			6		
		Appliance Input Rating in Thousands of Btu/h											
		FAN		NAT	FAN		NAT	FAN		NAT	FAN		NAT
		Min	Max	Max	Min	Max	Max	Min	Max	Max	Min	Max	Max
6	1	NA	NA	21	NA	NA	39	NA	NA	66	179	191	100
	2	NA	NA	28	NA	NA	52	NA	NA	84	186	227	123
	3	NA	NA	34	NA	NA	61	134	153	97	193	258	142
8	1	NA	NA	21	NA	NA	40	NA	NA	68	195	208	103
	2	NA	NA	28	NA	NA	52	137	139	85	202	240	125
	3	NA	NA	34	NA	NA	62	143	156	98	210	264	145
10	1	NA	NA	22	NA	NA	41	130	151	70	202	225	106
	2	NA	NA	29	NA	NA	53	136	150	86	210	255	128
	3	NA	NA	34	97	102	62	143	166	99	217	277	147
15	1	NA	NA	23	NA	NA	43	129	151	73	199	271	112
	2	NA	NA	30	92	103	54	135	170	88	207	295	132
	3	NA	NA	34	96	112	63	141	185	101	215	315	151
20	1	NA	NA	23	87	99	45	128	167	76	197	303	117
	2	NA	NA	30	91	111	55	134	185	90	205	325	136
	3	NA	NA	35	96	119	64	140	199	103	213	343	154
30	1	NA	NA	24	86	108	47	126	187	80	193	347	124
	2	NA	NA	31	91	119	57	132	203	93	201	366	142
	3	NA	NA	35	95	127	65	138	216	105	209	381	160
50	1	NA	NA	24	85	113	50	124	204	87	188	392	139
	2	NA	NA	31	89	123	60	130	218	100	196	408	158
	3	NA	NA	35	94	131	68	136	231	112	205	422	176

Vent Height H (ft)	Connector Rise R (ft)	7			8			9			10		
		Appliance Input Rating in Thousands of Btu/h											
		FAN		NAT	FAN		NAT	FAN		NAT	FAN		NAT
		Min	Max	Max	Min	Max	Max	Min	Max	Max	Min	Max	Max
6	1	231	271	140	292	366	200	362	474	252	499	594	316
	2	239	321	172	301	432	231	373	557	299	509	696	376
	3	247	365	202	309	491	269	381	634	348	519	793	437
8	1	250	298	146	313	407	207	387	530	263	529	672	331
	2	258	343	177	323	465	238	397	607	309	540	766	391
	3	266	376	205	332	509	274	407	663	356	551	838	450
10	1	267	316	151	333	434	213	410	571	273	558	727	343
	2	276	358	181	343	489	244	420	640	317	569	813	403
	3	284	389	207	352	530	279	430	694	363	580	880	459
15	1	268	376	161	349	502	225	445	646	291	623	808	366
	2	277	411	189	359	548	256	456	706	334	634	884	424
	3	286	439	213	368	586	289	466	755	378	646	945	479
20	1	265	425	169	345	569	235	439	734	306	614	921	387
	2	274	455	195	355	610	266	450	787	348	627	986	443
	3	282	481	219	365	644	298	461	831	391	639	1,042	496
30	1	259	492	183	338	665	250	430	864	330	600	1,089	421
	2	269	518	205	348	699	282	442	908	372	613	1,145	473
	3	277	540	229	358	729	312	452	946	412	626	1,193	524
50	1	252	567	208	328	778	287	417	1,022	383	582	1,302	492
	2	262	588	230	339	806	320	429	1,058	425	596	1,346	545
	3	271	607	255	349	831	351	440	1,090	466	610	1,386	597

(continued)

HEAT

TABLE 6-17 (CONTINUED)

Vent Height H (ft)	Minimum Internal Area of Masonry Chimney Flue (sq in.)								
	12			19			28		
	Combined Appliance Input Rating in Thousands of Btu/h								
	FAN +FAN	FAN +NAT	NAT +NAT	FAN +FAN	FAN +NAT	NAT +NAT	FAN +FAN	FAN +NAT	NAT +NAT
6	NA	NA	25	NA	118	45	NA	176	71
8	NA	NA	28	NA	128	52	NA	190	81
10	NA	NA	31	NA	136	56	NA	205	89
15	NA	NA	36	NA	NA	66	NA	230	105
20	NA	NA	NA	NA	NA	74	NA	247	120
30	NA	NA	NA	NA	NA	NA	NA	NA	135
50	NA	NA	NA	NA	NA	NA	NA	NA	NA

Vent Height H (ft)	38			50			63			78		
	Combined Appliance Input Rating in Thousands of Btu/h											
	FAN +FAN	FAN +FAN	NAT +NAT	FAN +FAN	FAN +FAN	NAT +NAT	FAN +FAN	FAN +FAN	NAT +NAT	FAN +FAN	FAN +FAN	NAT +NAT
6	NA	255	102	NA	348	142	NA	455	187	NA	579	245
8	NA	276	118	NA	380	162	NA	497	217	NA	633	277
10	NA	295	129	NA	405	175	NA	532	234	771	680	300
15	NA	335	150	NA	400	210	677	602	280	866	772	365
20	NA	362	170	NA	503	240	765	661	321	947	849	415
30	NA	398	195	NA	558	275	808	739	377	1,052	957	490
50	NA	NA	NA	NA	612	325	NA	821	456	1,152	1,076	600

DIRECT VENTING

In direct vent installations, combustion air is drawn into a sealed firebox from outside the house through coaxial intake/exhaust pipes. This eliminates depressurization, resulting in a warmer, healthier home environment. Direct venting also eliminates the need to extend the exhaust vent through the roof, making installation easier and less expensive.

Thorough wall venting for Category II and Category IV appliances or other condensing types of venting cannot terminate over other equipment that could be damaged by the condensation or terminate over public walkways.

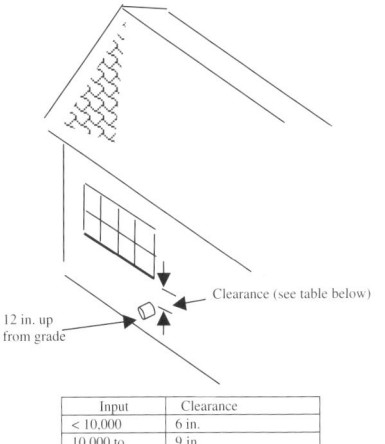

12 in. up from grade

Clearance (see table below)

Input	Clearance
< 10,000	6 in.
10,000 to 50,000	9 in.
> 50,000	12 in.

Figure 6-9 CLEARANCE FOR DIRECT VENT APPLIANCES

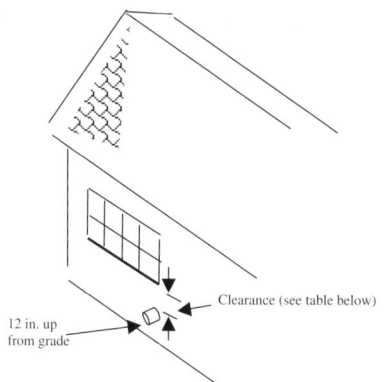

12 in. up from grade

Clearance (see table below)

Input	Clearance
< 10,000	6 in.
10,000 to 50,000	9 in.
> 50,000	12 in.

Figure 6-10 CLEARANCE FOR DIRECT VENT APPLIANCES

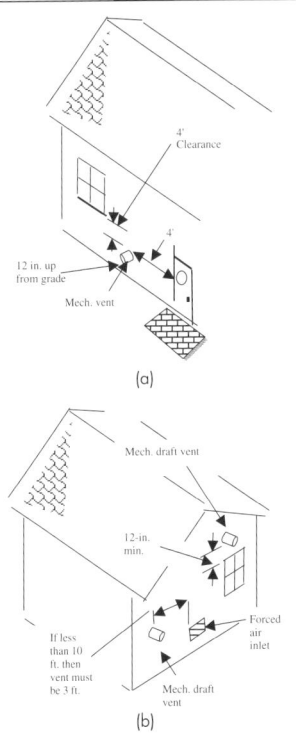

Figure 6-11   MECHANICAL DRAFT VENT LOCATIONS

HEAT

CLEARANCES FROM COMBUSTIONABLES

Clearances from Combustionable Materials for Connectors

TABLE 6-18   LISTED TYPE B VENT

EQUIPMENT	CLEARANCE REQUIRED
Equipment with draft hoods and those listed for use with Type B vents	As listed 6 in.
Residential Boilers, furnaces with listed gas conversion and draft hood	Not permitted
Residential appliances listed for use with Type L vents	Not permitted
Unlisted residential appliances with a drafthood	Not permitted
Residential and low-heat equipment other than ones listed above	

TABLE 6-19   LISTED TYPE L VENT

EQUIPMENT	CLEARANCE REQUIRED
Equipment with draft hoods and those listed for use with Type B vents.	As listed 6 in.
Residential Boilers, furnaces with listed gas conversion and draft hood	As listed
Residential appliances listed for use with Type L vents.	6 in.
Unlisted residential appliances with a draft hood	9 in.
Residential and low-heat equipment other than ones listed above.	

TABLE 6-20 SINGLE-WALL METAL PIPE

EQUIPMENT	CLEARANCE REQUIRED
Equipment with draft hoods and those listed for use with Type B vents.	6 in. 9 in.
Residential Boilers, furnaces with listed gas conversion and draft hood	9 in.
Residential appliances listed for use with Type L vents.	9 in.
Unlisted residential appliances with a draft hood	18 in.
Residential and low-heat equipment other than ones listedabove.	

Note: For factory built chimneys—as per installation instructions.

TABLE 6-21 THICKNESS FOR SINGLE-WALL METAL PIPE CONNECTORS

Diameter of Connector (in)	Galvanized Sheet Metal Gage Number	Minimum Thickness (in)
Less than 6	26	0.019
6–10	24	0.024
Over 10–16	22	0.029

Note: For oil or other solid fuels.

TABLE 6-22   CHIMNEY AND VENT CONNECTOR CLEARANCES TO COMBUSTIBLE MATERIAL

Type of Connector	Minimum Clearance (in)
Single-Wall Metal Pipe Connectors	
Oil and solid-fuel appliances	18
Oil appliances listed for use with type L vents	9
Type L Vent Piping Connectors	
Oil and solid-fuel appliances	9
Oil appliances listed for use with type L vents	3

DRAFT FANS

- Check fans and drivers for anchorage, alignment, and rotation.
- Check accessibility of lubrication fittings.
- Inspect dampers for operation in compliance with contract requirements.
- Inspect bearings for smoothness and overheating.
- Check vibration and vibration absorbing mounts.
- Inspect insulation application to induced draft fan.
- Examine safety control interlocks and sic-flow switches.

The following steps are critical to ensure adequate installation up-to-code condition of the fuel tank.

- ❏ Check for Underwriter's approval.
- ❏ Check tank capacity and calibration.
- ❏ See that tanks have the required openings and the means for proper anchorage.
- ❏ Check for tank heaters, when required.
- ❏ Examine paint coating and examine holiday testing.
- ❏ Check manufacturer instructions for proper installation.
- ❏ Inspect for capacity and for method of mounting, condensate, and vacuum return pumps.

MISCELLANEOUS FITTINGS AND EQUIPMENT

Inspect valves, drips, traps, coils, elements, convectors, radiators, etc. as they are brought on the job, to make sure that they are of the correct capacity and that they have been approved.

---

*For More Information on Fuel and Fuel Storage Tanks see "Up to Code Plumbing Systems."

TABLE **7.1**	LOCATION OF LP-GAS CONTAINERS	
CONTAINER CAPACITY (water gallons)	**MINIMUM SEPARATION BETWEEN CONTAINERS AND PUBLIC WAYS BUILDINGS, OR LOT LINES OF ADJOINING PROPERTY**	
	Mounded or underground Containers. (feet)	**Above-ground Container (feet)**
Less than 125	10	5
125 to 250	10	10
251 to 500	10	10
501 to 2,000	10	25
2,001 to 30,000	50	50
30,001 to 70,000	50	75
70,001 to 90,000	50	100
90,001 to 120,000	50	125

1. Minimum distance for underground containers shall be measured from the pressure relief device and the filling or liquid-level gauge vent connection at the container, except that all parts of an underground container shall be 10 ft or more from a building or lot line of adjoining property which can be built upon.
2. For other than installations in which the overhanging structure is 50 ft or more above the

relief-valve discharge outlet. In applying the distance between buildings and ASME containers with a water capacity of 125 gallons or more, a minimum of 50 percent of this horizontal distance shall also apply to all portions of the building which project more than 5 ft from the building wall and which are higher than the relief-valve discharge outlet. This horizontal distance shall be measured from a point determined by projecting the outside edge of such overhanging structure vertically downward to grade or other level upon which the container is installed. Distances to the building wall shall not be less than those prescribed in this table.

3. Containers of less than a 125-gallon water capacity are allowed next to the building they serve when in compliance with local codes.

THINK SAFETY AT ALL TIMES

FUEL STOR.

CHAPTER 8
INSPECTIONS

PLANNING

❑ Check the codes, reference data and manufacturer's recommendations.
❑ Check with contractor for his or her detail layouts of equipment and piping, which are normally made to coordinate work of the various trades.
❑ Compare nameplate data, piping markings, etc. with requirements.
❑ Provide the proper spacing of equipment to ensure adequate room for piping, ductwork, and accessibility for maintenance and so that walls behind ductwork can be finished without duct removal. Check for adequate clearance for removal of air filters and strainers.
❑ Verify how the heating system fits into the total job.
❑ Be sure that sleeves of the correct size and material are properly located in floors and walls before they are built.

PIPING

❑ Inspect piping workmanship.
❑ Check storage and handling procedures.
❑ Inspect for the required type and size of pipe.
❑ Examine the cutting of construction to install piping. (See chapter on cutting and notching of wood framing members.)
❑ Require provisions for expansion and contraction, and proper anchorage of pipe.
❑ Check the installation of mechanical expansion joints. Do not remove spacers until expansion joints are ready to be installed.
❑ Verify that the pitch of the horizontal runs are correct.
❑ Check the position of branch connections.
❑ Be sure that required valves are installed in the correct positions.
❑ Check valve types (globe vs. Gate, etc.).
❑ Check the method and procedure of jointing pipes.
   (a) On threaded joints, check for the use of tapered threads. See that graphite and oil, or their equivalent, are applied to the threads.
   (b) On welded joints, check for compliance with approved welding procedures, inspect for defective welds; check type of

INSPECT

material of the welding rod; make sure
welders have been qualified.

(c) See that piping is properly supported
and aligned and that there is no strain
on joints.

❏ See that proper grade and alignment are
maintained and that proper fittings are
provided to eliminate air pockets and
restrictions.

❏ Check for air valves at all high points and at
the ends of mains. Check for drips and traps
at low points. Examine the lines to make
sure that condensate cannot accumulate in
the lines.

❏ Inspect for required floor, wall, and ceiling
plates.

❏ Check for type, size, material, and finish.

❏ Watch for the use and proper installation of
eccentric fittings.

❏ See that interconnecting piping between
boilers conforms to shop drawings and
ASME Code. Watch for adequate valves
and other special fittings. Cutoff valves shall
be provided to isolate each boiler from the
steam header.

❏ Be sure that lift fittings are provided where
the gravity flow of vacuum returns is inter-
rupted by a change to a higher elevation.

- ❏ Clean all supply and return lines before putting them into operation.
- ❏ Check whether contractor has cleaned all traps and strainers after pipe cleaning and before system operation.
- ❏ Check safety valve discharge pipe for number of ells (restriction).
- ❏ Check bent pipe for kinks, wrinkles, or other malformations. Also be certain that approved radius bends are not exceeded.

PIPE INSULATION

Proper piping insulation is critical to maintenance. Note the following recommendations for securing adequate piping insulation.

- ❏ Know locations of pipes required to be insulated.
- ❏ See that insulation has been approved.
- ❏ Check width and type of material and the spacing of bands.
- ❏ Be sure that all fittings except unions and flanges are insulated.
- ❏ Be sure that insulation is being correctly installed.
- ❏ Check for continuity of insulation through walls and floors.

❑ Check that proper thickness of insulation is being applied.
❑ On chilled-water and hot-water combination piping and boiler piping, check for vapor seal requirement.

Systems must be examined as follows for high-quality installation, performance, and longevity. Note steps to complete to ensure satisfaction.

HOT WATER SYSTEMS

❑ Note the installation of balancing valves or orifices in the return connection of each radiator or heating device.
❑ See that contractor balances the system as required by plans or specifications.
❑ Ensure that threaded openings are provided on converters. See that safety devices and temperature controls are furnished and are in working order. Check coil for tightness and clearance for its removal. Note drain pipe to outside atmosphere or floor drain from blow-off safety valves.
❑ Check for automatic and manual vents.
❑ Examine expansion tanks for size, conformance to code, protective paint coating, insulation, water level gage, drain, and air charging valves.

High Temperature Hot Water

❏ Check pumps for:
   (a) Leveling, alignment, and stability on foundation
   (b) Lubrication
   (c) Seals for leaks
   (d) Packing adjustment and type
   (e) Pressure retention
   (f) Correct rotation
   (g) Seal coolant service installed.
❏ Ensure that radiant heating coils are accurately placed, firmly secured, and absolutely tight under a hydrostatic test pressure of one and one-half times the operating pressure prior to encasement in construction.

STEAM SYSTEMS

❏ Know details of the type of system required.
❏ Check the operation of supply valves to radiator and convector.
❏ Check radiator run-out for pitch.

Hot Air Heating

❏ Ensure that the contractor follows NFPA criteria for installation of oil or gas equipment.

❏ At no time will manufacturers instructions be changed and/or deviated from original plan.
❏ Be sure that return air has free passage to heater unit.
❏ Note damper setting balance of the flow of air.
❏ Check that flexible connections have been installed between the furnace and duct system.

HEATING AND VENTILATING UNITS

❏ Require that all component parts operate satisfactorily.
❏ Note access doors for tightness and clearance.
❏ Determine that noise level is within acceptable limits.
❏ Check flexible pipe connections and/or vibration eliminators.
❏ Check rotation.

Unit Heaters

(a) Check:
- Clearances
- Controls
- Air distribution
- Noise level
- Rotation

Controls

❏ Be sure that the controls are provided, as specified, that they are properly hooked up, and that they will perform the required operation.
❏ Has the required working space been provided?
❏ Has the required walkway to the working space been provided such as in attics?
❏ Working area requires a light.

BOILER SPECIALTIES INSPECTION

❏ Verify all trimmings (e.g., water column, steam gage, safety valves, blowoff valves, nonreturn valves). Stop and check feed valves and vent valves for type and size. Inspect for installation and setting.
❏ Check that safety valve discharge piping is anchored so that it does not impose a strain on the valve.

BOILER FEEDWATER

Check the following:

❏ Type of water treatment for:
  • Water available.
  • Pressures and temperatures to be obtained in boiler.
  • Materials and installation.

- ❏ Scales, proportioning devices, and mixing valves for accuracy and operation.
- ❏ Installation of tanks and piping for types of material and supports, workmanship, and conformance with contract requirements.
- ❏ Pressure tanks for conformance with the applicable codes and ASME stamp.
- ❏ Control apparatus for the installation and operation of all components. (Check should be done by the manufacturer's service engineer.) Refer to job specifications for necessary tests and reports required, and determine from service engineer the sequence of testing.
- ❏ Open heaters for the installation of pans, trays, plates, sprays, and other internal parts, as well as for the setting of the control for water level in storage compartment.
- ❏ Heater vent operation and that the heater reduces the oxygen content in the water to the specified amounts before acceptance. (Checking should be done by manufacturer's service engineer.)
- ❏ Closed heaters for compliance with code governing unfired pressure vessels. Ensure that clearance is provided for the removal of tubes. Evaluate performance.
- ❏ Thermometers and gages are accurate and are operating efficiently.

TURBINES

- ❏ Inspect equipment for the pressures and temperatures to be applied. Compare with approved shop drawings.
- ❏ Examine all drains, drips, leakoffs, relief valves, and other required safety devices for operation.
- ❏ Ensure that turbines are firmly secured to foundation, are accurately aligned with driven equipment, and operate without vibration.
- ❏ Check that piping is installed to impose no strain on turbine connections.
- ❏ Verify that provision is made for expansion when aligning couplings.
- ❏ Be certain that field-assembled turbines are installed by the manufacturer's erectors only
- ❏ Reduction gears must mesh perfectly and operate smoothly, without noise or vibration. Check dwelling after turbines and gears are in perfect alignment.
- ❏ Evaluate the operation of governors.
- ❏ Check capacity and steam consumption under various load conditions.

COMBUSTION CONTROLS

❏ Inspect equipment for type, capacity, installation, and operation.

❏ Be sure that operating devices are firmly secured to floor, foundations, or other supports and that they operate freely. They should have sufficient power to easily perform their duties.

❏ Check the location and stability of sleeves in setting walls, ducts or breechings for draft piping, thermometers, and gages.

❏ Pipe, tubing, and wiring should run neatly and parallel to the lines of the building or structure. They should be firmly secured and have proper pitch.

❏ See that draft piping is provided with a means for removing accumulations of ash and soot.

❏ Verify the operation of safety controls.

❏ Check that a flame-sensing device is installed in position to sense both pilot and main flame.

❏ Determine that instrument panels are firmly anchored and set plumb. Be sure that wiring, tubing, and piping are neatly arranged in rear of panel. See that nameplates, indicating the function of each instrument, are mounted on the face of the pane.

❏ Secure a written statement from the manufacturer's representative to the effect that all equipment of the control system is properly installed and in perfect operating condition before acceptance.

THINK SAFETY AT ALL TIMES

CHAPTER 9
SUPPLY, RETURN, AND COMBUSTIONABLE AIR

AIR TYPES

Return Air

Return air can be diluted with outdoor air. Openings shall not be less than 2 sq. in. per 1000 btu/h input rating of the furnace (for warm air furnace).

Return air for heat pumps and a central ac unit require a minimum of 6 in.^{2}/1000 btu/h nominal cooling output.

Prohibited Factors

- **Return air** for warm air furnaces from garages, bathrooms, kitchens, and other dwelling areas.
- Outdoor air taken from within 10 ft from either appliance or plumbing exhaust vents that are located less than 3 ft above the inlet.
- **Outdoor** inlets not covered with ¼ mesh with openings less than ½ in.

TABLE 9.1 RETURN/SUPPLY DUCT SIZING FOR HEAT PUMPS

Btu/hr	50K	60K	80K	100K	125K
Sq. in.	300	360	480	600	750
Round/Dia	20″	22″	26″	28″	32″

Heat pumps req. 6 sq. in. per 1000 Btu/hr—Check Manufacturers Instructions

K = 1000

Supply Air

Supply air requires <u>2 sq. in. per 1000 btu/h</u> for warm air furnaces.

For central air and heat pumps supply air shall be a <u>minimum 6 sq. in. per 1000 btu/h cooling</u> output.

Combustible Air

Combustible air does not apply to **direct vent** appliances, listed and labeled appliances, and domestic clothes dryers.

Buildings of unusually tight construction require combustible air from outside.

Buildings of ordinary tightness require a volume of <u>50 cu. ft. per 1000 btu/h input.</u>

Air requirements needed for other systems such as exhaust fans, fireplaces, and clothes dryers need to be factored in when computing the required air space.

AIR
SUPPLY

Openings for Air (adjusted net free passage)

- Metal Louvers: 75% of gross area
- Wood Louvers: 25% of gross area

AIR FROM INSIDE A BUILDING

Each opening shall have 1 sq. in per 1000 btu/hr, but not less than 100 sq. in.

If area does not meet the 50 sq. ft per 1000 btu/hr, then there shall be two openings to adjacent areas: one 12 in. from the top of the space and one 12 in. from the bottom.

AIR FROM OUTDOORS

- ❑ Requires two openings, one 12 in. from top of enclosure and one 12 in. from the bottom.
- ❑ Can be connected to crawl or attic if they are ventilated.
- ❑ Vertical ducts must have as a minimum 1 sq. in. per 4000 btu/hr of all appliances in the space.
- ❑ Horizontal ducts must have as a minimum 1 sq. in. per 2000 btu/hr.
- ❑ Ducts in attic must extend at least 6 in. above the ceiling joist and insulation.
- ❑ Ducts in attic shall not be screened.

❏ Ducts supplying combustible air from under floor areas require <u>twice</u> the required air opening.
❏ All outside openings require a corrosion-resistant screen with openings at least ¼ in. and not larger than ½ in.

COMBUSTION AND VENT AIR

All Air from Inside the Building

(a) A confined space with two openings:
- Each opening to have a minimum free area of not less than 1 sq. in. per 1000 btu of the total gas input of all the units together
- But not less than 100 sq. in.
- One 12 in. from the top, one 12 in. from the bottom
- Minimum dimension of at least 3 in.

All Air from Outdoors

(a) Two-Opening Method:
- Ducts must be same size as opening.
- Opening must be a minimum of 3 in.
- One 12 in. from the top, one 12 in. from the bottom

AIR SUPPLY

- If a vertical duct is used:
  —1 sq. in. of free area for each 4000 btu of all the units combined
- If a horizontal duct is used:
  —All dimensions the same as the vertical except the free opening shall be 1 sq. in. per 2000 of the total units in the space

(b) One-Opening Method
- 12 in. from the top of the enclosure
- 1-in. clearance from the sides and back
- 6-in. clearance from the front of the appliance
- 1 sq. in. per 3000 btu of total input of all the units.
- Not less than the sum of all vent connections

VENTILATION OPENINGS (CRAWLSPACES)

Ventilation openings shall be covered with any of the following materials, provided that the least dimension of the covering shall not exceed ¼ in. (6 mm):

- Perforated sheet metal plates not less than 0.070 in. thick
- Expanded sheet metal plates not less than 0.047 in. thick

- Extruded load-bearing vents
- Cast-iron grills or gratings
- Hardware cloth of 0.035 in. (0.89 mm) wire or heavier.
- Corrosion-resistant wire mesh, with the least dimension not exceeding ⅛ in. (3.2 mm)

Note: No gas or plumbing waste cleanouts can be located or pass through plenums.

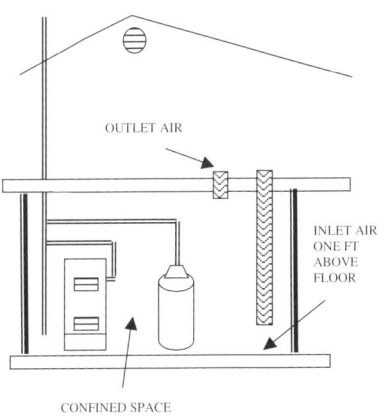

OUTLET AIR

INLET AIR
ONE FT
ABOVE
FLOOR

CONFINED SPACE

Figure 9.1 APPLIANCES IN A CONFINED SPACE WITH ALL AIR TAKEN FROM VENTED ATTIC

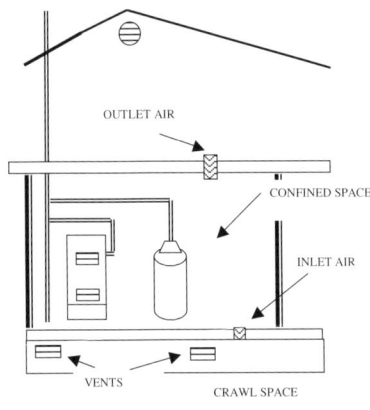

OUTLET AIR

CONFINED SPACE

INLET AIR

VENTS

CRAWL SPACE

Figure 9.2 APPLIANCES IN A CONFINED SPACE WITH INLET AIR TAKEN FROM THE CRAWL AND VENTED TO THE ATTIC

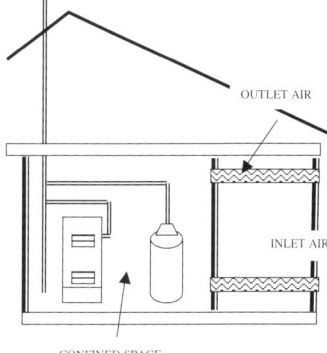

CONFINED SPACE

OUTLET AIR

INLET AIR

Figure 9.3   APPLIANCES IN A CONFINED SPACE WITH ALL AIR TAKEN FROM OUTSIDE.

TABLE 9.2 REQUIRED OPENING SIZES PER CODES BASED ON TOTAL BTU/H

All Air from Inside the Building (2 openings)

Input BTU per hr (all in the area)	1 sq.in. per 1,000 BTU Each opening (in)
8,000	8
10,000	10
20,000	20
30,000	30
40,000	40
50,000	50
60,000	60
70,000	70
80,000	80
90,000	90
100,000	100
110,000	110
120,000	120
130,000	130
140,000	140
150,000	150
160,000	160
170,000	170
180,000	180
190,000	190

All Air from Outside the Building (2 openings)	
Input BTU per hr (all in the area)	1 sq.in. per 4,000 BTU Each opening (in)
8,000	2
10,000	2.5
20,000	5
30,000	7.5
40,000	10
50,000	12.5
60,000	15
70,000	17.5
80,000	20
90,000	22.5
100,000	25
110,000	27.5
120,000	30
130,000	32.5
140,000	35
150,000	37.5
160,000	40
170,000	42.5
180,000	45
190,000	47.5

AIR
SUPPLY

TABLE 9.3 ALL AIR FROM OUTSIDE THE BUILDING (2 OPENINGS) FOR HORIZONTAL RUN DUCTS

Input BTU per hr (all in the area)	1 sq.in. per 2,000 BTU Each opening (in)
8,000	4
10,000	5
20,000	10
30,000	15
40,000	20
50,000	25
60,000	30
70,000	35
80,000	40
90,000	45
100,000	50
110,000	55
120,000	60
130,000	65
140,000	70
150,000	75
160,000	80
170,000	85
180,000	90
190,000	95

VENTS

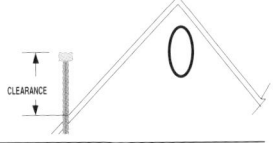

Roof slope	Clearance
Flat–6/12	1′–0″
6/12–7/12	1′–3″
7/12–8/12	1′–6″
8/12–9/12	2′–0″
9/12–10/12	2′–6″
10/12–11/12	3′–3″
11/12–12/12	4′–0″

Figure 9.4   VENT TERMINATIONS FOR LISTED CAPS, 12 IN. OR LESS AND AT LEAST 8 FT FROM VERTICAL WALL

- Metal ducts clearance to ground at least 6 in.
- Ducts in concrete—minimum 2-in. coverage.
- Joints must be taped or have gaskets (airtight)
- Connector requires a minimum of ¼ in. per ft.
- Supported with 1½-in. straps every 4 ft.
- Insulated if in unconditioned areas.

Types of Vents

Type B

For approved gas units with draft hoods and all other category I type appliances.

Type BW

For approved wall furnaces labeled for BW vents

Type L

For approved oil burning appliances and gas appliances approved for type B vents.

Venting Connections

❑ Examine the size and construction of stacks and flues.

- ❏ Check the clearance space between stacks, flues, and adjacent building materials.
- ❏ Inspect the method of supporting and anchoring all smoke connections.
- ❏ See that cleanout is provided which will allow cleaning of the entire smoke connection without dismantling.

VENTING GAS APPLIANCES

Vents that pass through insulation require a shield of at least 26-gage sheet metal. In attics the shield is to extend at least 2 in. above the level of insulation.

Mechanical draft vents must be at least 7 ft above grade if located adjacent public walkways.

MASONRY CHIMNEYS

Requirements/Specifications

1. Chimneys are to be lined with approved clay flue lining, a listed liner, or other approved material that will resist corrosion, erosion, and weakening from the vent gases at temperatures up to 1800°F.
2. If liners are installed in existing chimney structures (Type B), then the connection must

be marked not to allow any solid or liquid burning appliances to be attached.

3. Chimneys are to extend at least 5 ft above the highest connected equipment draft hood outlet or flue collar.

4. Sizing of Chimneys

   (a) The effective area of a chimney serving listed appliances with draft hood, Category I appliance listed for use with Type B vents are to be sized using Tables 6-22 to 6-29.

   (b) An exception to using the table for sizing for a single appliance with a draft hood, the vent connector and chimney flue will have an area not less than the appliance flue collar or draft hood and not greater than 7 times the draft hood outlet.

   (c) When two appliances with draft hoods are connected, the flue area must be the same as the largest appliance outlet plus 50 percent of the smaller appliance outlet and not greater than 7 times the smaller draft hood outlet.

Example

Largest outlet + 50% of smallest < chimney flue size < 7 × the smallest

5. Chimneys previously used for liquid or solid fuels and now using fuel gas must have a cleanout. It is to be located so that its upper edge is 6 in. below the lowest edge of the lowest inlet opening.
6. Offsets of Type B or Type L are not to exceed 45° from the vertical; however, one angle must not exceed 60°. The horizontal distance of a vent plus the connector vent shall not exceed 75 percent of the vertical height of the vent.

Chimney Venting

- Appliances installed in earthquake-prone areas need to be anchored to prevent lateral movement.
- Appliances that generate any type of ignitable source must have that source a minimum of 18 in. above the floor.
- Chimney shall extend at least 5 ft above the highest connected equipment draft hood.
- Open burner appliances such as a cooking range do not need automatic shut off device for use if the pilot light fails.
- Forced air furnaces need a control limit that prevents outlet air from exceeding 250°.
- Electric heater ducts need a limit of 200°F.

TABLE 9.4 FREE OPENING AREA OF CHIMNEY DAMPER FOR VENTING FLUE GASES FROM UNLISTED DECORATIVE APPLIANCES FOR INSTALLATION IN VENTED FIREPLACES

CHIMNEY HEIGHT (feet)	MINIMUM PERMANENT FREE OPENING square inches						
	8	13	20	29	39	51	64
	Appliance input rating (Btu per hour)						
6	7,800	14,000	23,200	34,000	46,400	62,400	80,000
8	8,400	15,200	25,200	37,000	50,400	68,000	86,000
10	9,000	16,800	27,600	40,400	55,800	74,400	96,400
15	9,800	18,200	30,200	44,600	62,400	84,000	108,800
20	10,600	20,200	32,600	50,400	68,400	94,000	122,200
30	11,200	21,600	36,600	55,200	76,800	105,800	138,600

THINK SAFETY AT ALL TIMES

GENERAL SPECIFICATIONS AND REQUIREMENTS

❑ Need to be installed in such a manner as to allow their replacement and/or repair.
❑ Need a power receptacle within 25 ft.
❑ Attic installation needs to provide a minimum of 22 ft wide × 30 ft long. Also requires a passage to the area.
❑ Need a permanent light switch and outlet with the switch located at the passageway opening.
❑ Units installed on the ground, in the crawlspace need to be on a slab or masonry units at least 3 ft high and level.
❑ Units suspended from the floor need a clearance of at least 6 in. from the ground.
❑ If equipment is installed in an excavation, it will have at least 6-in. clearance on the bottom and 12 in. on the sides, the control side still needs a minimum of 30-in. clearance.
❑ If the excavation exceeds 12 in., it needs concrete or masonry units installed to a height of 4 in. above the ground height.

❑ Fuel burning furnaces will not be installed in rooms used for storage.

CARBON MONOXIDE

A biproduct of burning fuels, carbon monoxide is a deadly gas that kills in a matter of minutes. Carbon monoxide is inhaled into the lungs where it attaches to the hemoglobin and moves into the blood stream. Hemoglobin has a much greater attraction to CO than oxygen and forms a very strong bond with the CO. Hemoglobin that has bonded with CO is unable to carry oxygen to the body, so its victim suffers from a lack of oxygen, thus actually suffocating.

Symptoms of CO Poisoning

- Flu-like symptoms
- Tightness across forehead
- Headache
- Partial loss of muscular control
- Increase pulse and respiration
- Dizziness
- Weakness
- Nausea

Allowable CO Levels in Appliances

Range	800PPM
Dryer	400PPM
Water heater	200PPM
Unvented heater	200PPM
Vented space heater	200PPM

Suspect CO When:

- Condensation on windows
- Plants are dying
- Odor of aldehyde
- Owners complain of headaches, nausea

APPLIANCE TYPES

Category I

Appliances that operate with a nonpositive vent connection pressure and with a flue gas temperature of at least 140°F above its dewpoint.

Category II

Appliances that operate with a nonpositive vent connection pressure and with a flue gas temperature less than 140°F above its dewpoint.

Category III

Appliances that operate with a positive vent pressure and with a flue gas temperature at least 140°F above its dewpoint.

Category IV

Appliances that operate with a positive vent pressure and with a flue gas temperature less than 140°F above its dewpoint.

Painting

- ❑ See that equipment contains the correct finish. Watch for abrasions.
- ❑ Watch for miscellaneous ferrous metal items that are not primed.
- ❑ Require finish painting as specified.
- ❑ Identify all pipe runs as specified.

TESTING

Witness that all required tests of heating equipment are accurately recorded. See that tests are performed by the manufacturer's representatives where required. Check tests and verify that tests meet all requirements before acceptance. Report unsatisfactory test results to the supervisor.

OPERATING INSTRUCTIONS AND GUARANTIES

❑ See that equipment guaranties and instructions for the operation of equipment are furnished for handover to the client.

❑ Notify the site supervisor of the readiness of the construction for test and subsequent operation for instructing personnel. Video-taping is an excellent tool.

THINK SAFETY AT ALL TIMES

H/C
EQUIP.

CHAPTER 11
VENTILATING, AIR SUPPLY, AND DISTRIBUTION SYSTEM

INTRODUCTION

Because the ventilating system is largely dependent on associated equipment, the Inspector must closely coordinate this chapter with all the additional chapters in this inspection manual.

TABLE 11-1	ASHRAE RECOMMENDED DESIGN CRITERIA		
		Inside Design Conditions	
Area		**Winter**	**Summer**
Dining and entertainment	Cafeterias	70 to 74°F 20 to 30% rh	78°F 50% rh
	Restaurants	70 to 74°F 20 to 30% rh	74 to 78°F 33 to 60% rh
	Bars	70 to 74°F 20 to 30% rh	74 to 78°F 30 to 60% rh
	Nightclubs	70 to 74°F 20 to 30% rh	74 to 78°F 30 to 60% rh
	Kitchen	70 to 74°F	85 to 88°F
Office building		70 to 74°F 20 to 30% rh	74 to 78°F 50 to 60% rh
Libraries and museums	Average	61 to 72°F 40 to 33% rh	
	Archival	Special considerations	

Area		Inside Design Conditions	
		Winter	Summer
Bowling		70 to 74°F 20 to 30% rh	73 to 78°F 23 to 33% rh
Communi-cation centers	Telephone Terminal Rooms	72 to 78°F 40 to 30% rh	72 to 78°F 40 to 30% rh
	Teletype Centers	70 to 74°F 40 to 30% rh	74 to 78°F 43 to 33% rh
	Radio and Television Stations	74 to 78°F 30 to 40% rh	74 to 78°F 40 to 33% rh
Transpor-tation centers	Airport Terminals	70 to 74°F 20 to 30% rh	74 to 78°F 30 to 60% rh
	Ship Docks	70 to 74°F 20 to 30% rh	74 to 78°F 30 to 60% rh
	Bus Terminal	70 to 74°F 20 to 30% rh	74 to 78°F 30 to 60% rh
	Garages	40 to 33°F	80 to 100°F
Warehouses		Inside design often depend on the materials stored	

(continued)

TABLE **11.1**	(CONTINUED)	Inside Design Conditions	
Area		**Air Movement**	**Air Changes (per hr)**
Dining and entertainment	Cafeterias	50 fpm at 6 ft above floor	12 to 15
	Restaurants	25 to 30 fpm	8 to 12
	Bars	30 fpm at 6 ft. above floor	15 to 20
	Nightclubs	below 23 fpm at 3 ft. above floor	20 to 30
	Kitchen	30 to 50 fpm	12 to 15
Office Building		25 to 45 fpm	4 to 10
Libraries and museums	Average	below 23 fpm	8 to 12
	Archival	below 23 fpm	8 to 12
Bowling		50 fpm at 6 ft. above floor	10 to 13
Communication centers	Telephone Terminal Rooms	25 to 30 fpm	8 to 20
	Teletype Centers	25 to 30 fpm	8 to 20
	Radio and Television Stations	below 23 fpm at 12 ft. above, floor	15 to 40

Area		Inside Design Conditions	
		Air Movement	Air Changes (per hr)
Transportation centers	Airport Terminals	25 to 30 fpm at 6 ft. above floor	8 to 12
	Ship Docks	23 to 30 fpm 6 ft. above floor	8 to 12
	Bus Terminal	25 to 30 fpm at 6 ft. above floor	8 to 12
	Garages	30 to 75 fpm	4 to 6 (NFPA)
Warehouses			1 to 4

Note: (rh = relative humidity level)

EQUIPMENT

Introduction

The Inspector determines that all equipment is approved well in advance of its actual need on the job.

❑ Check all equipment delivered to the site for conformance with approved shop drawings. Make sure the necessary rating and test certificates have been furnished.

❑ Closely examine material for any damages. Minor abrasions or rust spots must be cleaned and repainted to match original paint in appearance and in quality. Reject damaged items.

❑ Be certain that approved vibration-isolators and flexible connections will be furnished as specified.

❑ Examine the mounting of each piece of equipment for secure installation. Check codes for maximum spacing of supports.

❑ Check equipment for excess noise and vibration.

❑ Do not use dissimilar materials, especially screws, fasteners and flashings with different equipment bases and housing materials.

Fans and Air Handling Units

❑ Check rotation of the fan before permanent power connection is made.

❑ Check method of drive. If belt driven, check the means provided to adjust the motor.

❑ Check the type of motor enclosure.

❑ See that specified seals, sleeves, and bearings are provided, and when lubricating type bearings are allowed provide accessibility for lubricating without dismantling fan or disconnecting duct.

- ❑ Provide a fire-safety switch on return air ducts of circulation systems.
- ❑ Check for pulley and belt alignment.
- ❑ See that adequate guards are provided for rotating equipment and belts.
- ❑ Check for installation of smoke detectors when required.

Power Roof Ventilators

- ❑ Provide service accessibility.
- ❑ Flashing at curbs must be water tight.
- ❑ Discharged air is not to be directed toward air intakes.
- ❑ Check for required disconnect switch—should be visible and properly marked.

Gravity Ventilators

- ❑ Examine installation for rigidity and weather tightness.
- ❑ Make sure units are oiled and properly adjusted.
- ❑ Check the actual freedom of rotation of the blades.

Ventilator Calculations

To establish the required number of ventilators needed for any area, first calculate the total number of cubic feet in the area to be ventilated. Then follow the instructions below.

If You Know:

- **Required Air Changes** Divide the total cubic feet by the number of air changes required per minute. This will give you the number of CFM required to reach the correct air changes. On the CFM ratings chart at the right, find the RV that most closely reaches the correct CFM (use the lowest CFM rating so that this will be the minimum attained).
- **Size of Rotary Requested:** Divide minimum CFM rating into total cubic feet. This will tell you how many ventilators are needed to totally ventilate the area each minute. (This number can then be used to calculate the number of minutes within which you want a 100% air change to take place.)

TABLE 11-2   AIR CHANGE PER SIZE OF VENTILATOR

Cubic Feet per Minute (CFM)

WIND SPEED	TEMP. DIFFER.	6"	8"	10"
4MPH	10	81	144	224
	20	83	149	232
	30	88	149	244
8MPH	10	144	257	400
	20	146	261	407
	30	146	264	410
10MPH	10	178	317	494
	20	182	325	506
	30	188	332	517

TABLE 11-3   AIR CHANGE PER SIZE OF VENTILATOR

Cubic Feet per Minute (CFM)

WIND SPEED	TEMP. DIFFER.	12"	14"	18"
4MPH	10	323	440	721
	20	334	455	752
	30	352	479	793
8MPH	10	576	788	1299
	20	586	798	1320
	30	591	806	1333
10MPH	10	711	970	1604
	20	729	994	1844
	30	748	1016	1678

Dampers

- ❏ Backdraft dampers should be installed for each exhaust fan.
- ❏ Check the actual operation of the dampers. See that dampers do not rattle and that felt strips are provided for backdraft dampers.
- ❏ Ensure that a separate frame is provided in openings on which the dampers will be mounted.
- ❏ Check for correct installation of fire dampers in accordance with SMACNA Fire Damper Book.

Filters

- ❏ Make sure the proper type of filter is furnished and installed.
- ❏ Check thickness and method of mounting and supporting.
- ❏ Provide proper amount of adhesive and washing tank for viscous medium type filters.
- ❏ Inspect sealing strips.
- ❏ Provide accessibility for removal and replacement of filters.
- ❏ Ensure that air stream is distributed uniformly over all filter areas.

- ❏ Observe electrostatic-type filters for operation of warning lights and door interlocks. Check ionizers for loose wires, sparking, and free access.
- ❏ Inspect automatic sprays for complete washing and spray coverage.
- ❏ On traveling screen type filters, note the operation of screen and oil charge.
- ❏ On renewable roll media type filters inspect:

  - Tracking of roll
  - Media runout switch
  - Timer setting
  - Static pressure control
  - Tension on media

- ❏ See that clean filters are installed upon completion of final tests.
- ❏ Check specifications regarding requirements for spare filters. This requirement is sometimes expressed as a percentage of the total of each kind required. Check on the transfer of the spares to the operating agency.

Screens

- ❏ Provide bird or insect screens if required.
- ❏ Check fabric material and installation of dissimilar materials.
- ❏ Check mesh size.

DIFFUSERS, REGISTERS, AND GRILLES

- ❏ See that the contractor furnishes a schedule showing all air inlets and outlets.
- ❏ Inspect diffusers and registers for accessible volume control operator.
- ❏ Examine specification and installation for integral antismudge rings for diffusers.
- ❏ Check for loose or bent vanes.
- ❏ Inspect each item for fit, and see that sponge-rubber gaskets are provided when required.
- ❏ Inspect for the proper operation of registers, dampers, and grille directional controls.

Balancing and Testing

It is important to check for any required certification of the HVAC test and balance.

Cleaning and Adjusting

- ❏ All ducts, plenums and casings must be thoroughly cleaned of debris and blown free of small particles and dust before supply outlets are installed.
- ❏ Clean equipment of oil, dust, dirt, and paint spots.
- ❏ Replace sectional throw-away filters after ductwork is blown out and cleaned.
- ❏ Lubricate all bearings.
- ❏ Check tension on all belts and the adjustment of fan pulleys.
- ❏ Check that all fan and belt guards are in place.
- ❏ Install temporary filters for testing purposes.

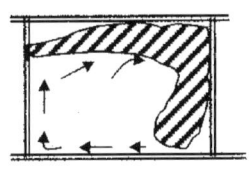

Cooling

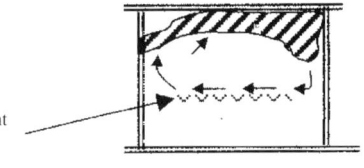

Stagnant

Heating

**Figure 11-1   AIR MOVEMENT FROM TYPE A OUTLETS
(HIGH SIDE WALL)**

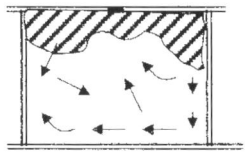

Cooling

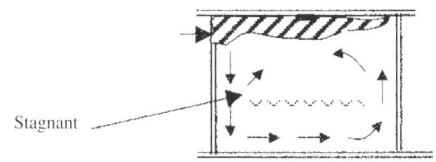

Stagnant

Heating

Figure 11-2   AIR MOVEMENT FROM TYPE A OUTLETS (CEILING)

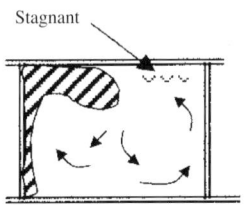

Stagnant

Cooling

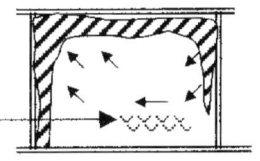

Stagnant

Heating

Figure 11-3   AIR MOVEMENT FROM TYPE B OUTLETS

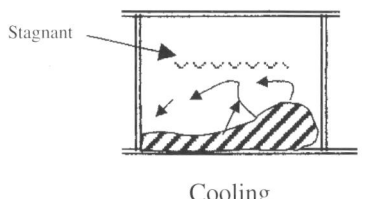

Stagnant

Cooling

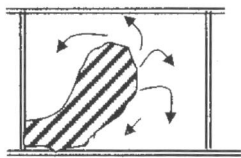

Heating

Figure 11-4   AIR MOVEMENT FROM TYPE D OUTLETS

Testing

- ❏ Before insulating the duct, test it for air tightness. Contractor must provide necessary equipment for airflow measurements and coefficients for registers and diffusers.
- ❏ Review contractor's method for recording test data, including comparison to the design airflows.
- ❏ Test each outlet for the amount of air quantities required.
- ❏ Final airflows must be recorded after all adjustments are made.
- ❏ If actual airflows result in objectional velocities or distribution, notify the Project Manager.
- ❏ Check all dampers for proper operation.

THINK SAFETY AT ALL TIMES

CHAPTER 12
REFRIGERATION AND AIR CONDITIONING

INTRODUCTION

This chapter covers Refrigeration and Air Conditioning for both the central and unitary type systems.

Because there is generally a duplication in the requirements for piping and ductwork for this subject and for plumbing and air handling, and since those areas have been covered in previous chapters, the Inspector must be very familiar with the piping section and the ductwork section.

PIPING

Refrigerator Piping

- ❑ Determine where copper or black steel will be used, and the type required.
- ❑ Make sure the piping and fittings have been approved.
- ❑ Check the method of installing piping.
- ❑ Make sure piping is stored as prescribed in specifications.

Water Piping

- ❑ Check the type of piping required for chilled water and condenser water systems.
- ❑ Determine the weight and class of the piping.
- ❑ Make sure the specified and approved piping, fittings, and jointing materials are being used.

Installation

- ❑ Check for defects in fabricating and installing piping. Watch specifically for workmanship, supports, and sleeves.
- ❑ Be especially careful to:
  - Make sure the specified solder is used. Check soldering of joints.
  - See that internal valve parts are removed from valves, and that valves are wet wrapped before soldering.
  - See that joints are thoroughly cleaned before soldering.
  - Check on the removal of excess flux and acid after joints are made.
- ❑ Make sure the proper type flexible connections are installed in the required locations.
- ❑ See that unions or flanges are installed at all equipment, at control valves, and at other points that will facilitate maintenance.

- ❏ Check carefully for the proper slope of all lines. Ensure that the slope of refrigerant lines provides for the movement of the oil through the system.
- ❏ Check installation for improper configuration of piping. Make sure the installation conforms with the approved drawing. If there is any question about the requirement for the arrangement of piping and if there is no approved drawing, obtain the drawing before allowing the contractor to proceed.
- ❏ Make sure air vents are installed at high points in water lines, and that drains are installed at low points.
- ❏ Do not allow gate valves to be installed where globe valves are required.
- ❏ Be sure balancing cocks are installed as required to permit proper balancing.
- ❏ Do not install swing check valves in vertical lines with a downward flow of water.
- ❏ Check for the installation of such required items as pressure gages, thermal elements, thermometer wells, etc.
- ❏ Provide adequate number and type of hangers.
- ❏ Hangers on uninsulated copper pipe must be electrolytically coated or made of solid compatible nonferrous metals.

- ❏ Check for the installation of expansive type fittings where required. (all building expansion joints)
- ❏ Check for the proper installation of oil traps and double risers in refrigerant lines.
- ❏ Check valves for pressure setting and discharge locations.
- ❏ Be sure that the refrigerant system is evacuated prior to charging and accomplished according to job specifications.
- ❏ Make sure the system is charged with the required type and amount of refrigerant.
- ❏ See that the system is completely checked for leaks. Dry nitrogen should be used for pressure tests unless another system has been approved.
- ❏ Double check to see that there are no unnecessary oil traps.
- ❏ Vacuum should be broken by charging the system with dry refrigerant for which the system is designed.

Insulation

- ❏ Determine whether the material on the job has been approved for the particular piping being installed. Make sure insulation, vapor barriers, adhesives, and sealers are noncombustible or fire retardant as specified.

- ❑ Note that heating water piping is insulated differently from both chilled water piping and combination chilled and heated water piping.
- ❑ Check thickness of insulation and of vapor barrier finish.
- ❑ Determine that insulation jackets which are exposed to view are paintable.
- ❑ Examine the requirements for the insulation of flanges, fittings, and valves, and ensure compliance with the specifications.
- ❑ Check the lap and the sealing at joints.
- ❑ Be very careful to see that there are no breaks in the vapor barrier. Watch for later damages during construction.
- ❑ Check specification requirements for extending insulation through sleeves in walls, floors, and ceilings; chilled water lines inside cabinets of fan coil units should be insulated as required to prevent condensation from dripping on floor.
- ❑ Make sure that pipe hangers are installed over insulation. Metal shields must be provided between hanger ring and insulation. High-density insulation insert shall be installed with a length equal to the length of metal shield.
- ❑ Know the special requirements for insulation and jacketing of piping exposed to weather.

REFRIG./ AC

- ❏ Check for the neat termination and seal of insulation at the end of insulation.
- ❏ Check the installation, the width, and the spacing of the bands used on pipe jacketing.

EQUIPMENT

Introduction

- ❏ All equipment should be checked to see that it is approved before it is needed on the job. When equipment arrives on the job, it should be checked against shop drawing.
- ❏ During installation, the contractor's work should be checked against the contract plans and specifications, the approved shop drawing, and the manufacturer's recommendations. Be sure that no damaged equipment is installed.
- ❏ See that equipment is stored and protected until installed.
- ❏ Be sure that all refrigeration equipment is installed strictly in accordance with the code.
- ❏ Check on space requirements for equipment. Obtain an equipment room layout drawing and make sure that adequate clearances are provided for installation, mainte-

nance and operation. Are doors large enough to allow for equipment to be installed?

❑ Determine the need for access panels. A common error is the failure to provide the means for pulling condenser and chiller tubes.

❑ Check the type of motors on equipment, the type of motor starter, heaters in the motor starters, and voltage of motor.

❑ Make sure that all rotating parts, such as belts, chains, sheaves, shaft couplings, etc. are covered to protect personnel.

❑ Make sure all equipment is lubricated according to the manufacturer's instructions.

Condensers

❑ See that air flow is not obstructed and that wind deflectors are installed, if required, in air-cooled condensers.

❑ Inspect water-cooled condensers for leaks and proper flow.

❑ Check evaporative condensers for:
- Spray coverage
- Float valve operation without chatter
- Water level
- Fan rotation and speed

- Pump suction strainer
- Liquid discharge line carried full size to first elbow, with a 12-in. to 18-in. drop to receiver
- Mesh size of inlet screens
- Pan, casing, eliminators, fan corrosion protection, and complete drainage
- Provision for and adjustment of constant bleeding.

❑ For all-season air-cooled condensers, manufacturers recommended installation should be adhered to. Check project plans, specifications, and manufacturers recommended installation to see if condenser flooding or air volume control is required.

Reciprocating Compressors

Check for:

❑ Oil, suction, and discharge pressures.
❑ Shaft alignment on direct-driven machines.
❑ Operation of high pressurestat, low pressurestat, and oil pressure failure switch.
❑ Proper level viscosity of oil.
❑ Installation of required gages.
❑ Amount, correct type, and dryness of refrigerant charge.
❑ Pressure holding ability upon pump-down.

- ❑ Isolator deflection and compressor vibration.
- ❑ Suction strainer screen mesh, and removal of startup belts.
- ❑ Unloader action.
- ❑ Compressor speed.
- ❑ Belt tension and alignment.
- ❑ Motor amperage under maximum load.
- ❑ Refrigerant flood back and oil foaming.
- ❑ Cylinder head overheating.
- ❑ Rotation.
- ❑ Automatic oil heater in crank case. Heater should work during shutdown.
- ❑ Loops in refrigerant piping as loops will permit oil to become trapped.
- ❑ Damage of equipment compressor—should not be run during vacuum tests.
- ❑ Clearances from grade.
- ❑ Unit is level.

Centrifugal Compressors

Check for:

- ❑ Alignment of compressor, drive and gear box.
- ❑ Suction damper or inlet vane operation.
- ❑ Safety control circuit operation.
- ❑ Purge compressor operation.
- ❑ Float valve operation, if furnished.

❏ Oil pump and cooler operation.
❏ Noise and vibration.
❏ Required gages.

Receivers

Check for:

❏ Location, if installed on the outside of the building.
❏ Do not place in direct rays of sun.
❏ Relief valves are adequate size.
❏ ASME Stamp.
❏ Drain, purge valve, liquid level indication, and shut-off valves.

Water Chillers

❏ Examine water drains, vents, and correct pass arrangement in direct expansion type chillers.
❏ Inspect for freeze protection safety devices.
❏ Check strength of liquid bleed-off at bottom of flooded chillers. Check adjustment of level control.
❏ Check tubes and shell in brine chiller for type of material.

Evaporative Coolers

Inspect for adequate spray coverage, non-sagging media, water carrythrough, correct water level in sump, and lack of float valve "chatter."

Unit Coolers

Check:

- ❏ For corrosion-protected pan and casing.
- ❏ Water defrost units for spray coverage with no carryover.
- ❏ Electric defrost units, for cycle timing in accordance with the job conditions.
- ❏ Hot gas defrost, for suction pressures and refrigerant charge in accordance with manufacturer's instructions.
- ❏ Drainage during defrost cycle.
- ❏ Cycle timing.
- ❏ Check that drain lines are properly trapped on the warm end.

Refrigeration Specialties

Check:

- ❏ Superheat setting of expansion valves and that bulb and equalizer position is in accor-

dance with the manufacturer's recommendations.
❑ Solenoid valve for vertical stem, correct
direction of refrigerant flow, and manual
opener disengagement.
❑ For unobstructed view of sight glass.
❑ Operation of evaporator pressure regulator
under light load.
❑ Operation of hold-back value upon start-up.
❑ Float valves or switches mounted level and
at a height that will ensure correct liquid
level in the evapoess before opening of
refrigerant drier canisters.
❑ Air tightness before opening of refrigerant
drier canisters.
❑ Drier; if it is the replaceable type, piping
will be arranged to facilitate replacement—
three-valve bypass.
❑ Piping connections of liquid-suction heat
exchanger.
❑ That direct expansion coils are installed as
recommended by manufacturer.
❑ That pans of fan-coil units are protected
against corrosion.
❑ That drain pans are installed under all units,
or as needed, to collect condensate.

Package-Type Air Conditioners

Check:

❑ High-pressure cutout setting.
❑ Compressor hold-down bolts are (for shipping) removed.
❑ Drip pan is watertight and connected to open drain.
❑ Water regulator valve operation, if used.
❑ Installation of air filters and strainers.
❑ Operation of thermostat.
❑ Suction and discharge pressures of refrigeration compressors.

TABLE 12-1   PACKAGED TERMINAL AIR CONDITIONERS (PTAC) AND PACKAGED TERMINAL HEAT PUMPS (PTHP) ELECTRICALLY OPERATED, MINIMUM EFFICIENCY

Equipment Type	Rating Condition	Minimum Efficiency
PTAC and PTHP	95°F db outdoor air	$10.0 - (0.16 \times Cap/1000)$ EER
(cooling mode)	82°F db outdoor air	$12.2 - (0.20 \times Cap/1000)$ EER
PTHP (heating mode)	—	$2.9 - (0.026 \times Cap/1000)$ COP

"Cap" is the rated capacity of the product in Btu/h. If the unit's capacity is less than 7000 Btu/h, use 7000 Btu/h in the calculation.

If the unit's capacity is greater than 15,000 Btu/h, use 15,000 Btu/h in the calculation.

TABLE 12-2 WARM AIR FURNACES AND COMBINATION WARM AIR FURNACES/AIR CONDITIONING UNITS, HEATERS, MINIMUM EFFICIENCY

Equipment Type	Size	Rating Condition	Minimum Efficiency
Warm air furnace, gas-fired	<225,000 Btuh	—	78% AFUE or 80% E
	>225,000 Btuh	Maximum capacity	80%E
Warm air furnace, oil-fired	<225,000 Btuh	—	78% AFUE or 80% E
	>225,000 Btuh	Maximum capacity	81%E
Warm air duct furnace, gas-fired	All capacities	Maximum capacity	78%E
		Minimum capacity	75%E
Warm air unit heater, gas-fired	All capacities	Maximum capacity	78%E
		Minimum capacity	74%E
Warm air unit heater, oil-fired	All capacities	Maximum capacity	81%E
		Minimum capacity	81%E

Note: E = Thermal efficiency.

TABLE 12-3	HEAT PUMPS COOLING MODE	
Size	**Type**	**Minimum Efficiency**
Less than 65,000 Btu/h	Split System Single Package	10.0 SEER 9.7 SEER
65,000 Btu/h to less than 135,000 Btu/h	Split System and Single Package	8.9 EER
135,000 Btu/h to less than 240,000 Btu/h	Split System and Single Package	8.5 EER
240,000 Btu/h to less than 760,000 Btu/h	Split System and Single Package	8.5 EER
Less than 760,000 Btu/h	Split System and Single Package	8.2 EER

TABLE **12-4**	HEAT PUMPS HEATING MODE	
Size	Type	Minimum Efficiency
Less than 65,000 Btu/h	Split System	6.8 HSPF
	Single Package	6.6 HSPF
65,000 Btu/h to less than 135,000 Btu/h	47 F° db/43 F°db	3.0 COP
	17 F° db/15 F°db	2.0 COP
135,000 Btu/h to less than 240,000 Btu/h	47 F° db/43 F°db	2.9 COP
	17 F° db/15 F°db	2.0 COP
More than 240,000 Btu/h	47 F° db/43 F°db	2.9 COP
	17 F° db/15 F°db	2.0 COP

Humidifiers and Dehumidifiers

❏ Examine the humidifiers for the supported coil and corrosion-protected pan.
❏ Check refrigeration-type dehumidifiers for frosting of cooling coil and for water carry-over.
❏ Check absorption type dehumidifiers for:
  • Solution level and temperature controls.
  • No solution should carry over from eliminators.
  • Regenerator duct must be drained of specified material, and correctly sealed.

- Damper operation, cycle timing, evidence of "dusting" of the desiccant, and regenerative temperatures.

Absorption Refrigeration Machine

Check the following:

❑ Cleanliness of all parts during erection.
❑ Proper materials.
❑ Access for removing tubes from absorber-evaporator and generator-condenser.
❑ Control operation, especially high- and low-limit temperature cutouts or condenser water pump interlock.
❑ Operation of purge system.
❑ Unit to be fully charged with water and a nontoxic absorber after installation.
❑ Available services of a factory representative for charging, testing, starting the plant, and providing instruction.

Cooling Towers and Ponds

❑ Check mechanical-draft cooling towers for unobstructed air intake, fan rotation and speed, belt tension, stacked fill, and weather protection of motor. (Do not allow open fan motors when totally enclosed motors are specified.) Ensure that waterflow through the outlet does not form a vortex which draws air in with the water. Check operation of water temperature control and drainage devices.

❑ Observe spray ponds for evenness of sprays and for water drift.

❑ Ensure provision for an adjustment of constant bleed.

❑ See that mist eliminators are installed when specified.

❑ Ensure the installation of overflow and drain piping.

❑ See that the water is at an adequate level after operation, and that spray-pump operates.

❑ Check belt alignment and tension.

Pumps

❑ Ensure that manufacture nameplates, equipment, serial numbers, or code stamps are not covered or hidden from view after installation.

❑ Check for anchorage of pump in compliance with contract.

❑ Check alignment of pump with motor and piping.

❑ Make sure that all gages and meters are provided.

❑ See that eccentric reducers, in lieu of concentric reducers, are used in suction piping, and that the flat side is turned up.

❑ Check for adequate support of piping around pump.

❑ Be sure check valve is installed in discharge piping.

❑ Check pump packing. Make sure adequate packing is installed to allow gland take-up.

❑ Check for excess vibration and flexible piping connections if required.

❑ Make sure that the pump motor is weatherproof when specified, and that it is connected to rotate correctly.

❑ Recheck oil sumps after operation, if applicable.

INSULATION

❑ Check for:
- Proper insulation of chilled water pumps.
- Insulated converters and expansion tanks.
- Insulated condensate drain pans of air handling units.
- Protective finish over such items as pumps, converters, tanks, and fans.

REFRIG./
AC

- ❑ Ensure that all insulating materials have been approved, and that they are of the specified thickness.
- ❑ Check the method of attaching insulation to equipment.
- ❑ Make sure that specified reinforcing is provided in adhesive plaster finish.
- ❑ See that corner angle beads are installed at the specified corners.
- ❑ See that the adhesive finish coat has a smooth, pleasing finish.
- ❑ Check on the application of vapor barriers to see that they effectively seal out all moisture.

CONTROLS

- ❑ Review all control installations with approved control shop drawings to ensure that they are being installed in strict conformance with the drawings.
- ❑ See that dampers are mounted securely on rigid supports, and that the correct bearings are provided on the blade axles.
- ❑ Note damper motors while fan is on, and check linkage between damper and motor.
- ❑ Examine valve operations for tight closing.
- ❑ Examine electrical equipment for interlocking.

- ❏ Check on the installation of all required alarm bells.
- ❏ See that freeze-stats are installed as specified.
- ❏ Check for proper electrical current and voltage in the control system. Carefully check the operation of solenoid valves.
- ❏ Check that the air compressor location will permit tank drain operation, and check for cycle time with all operating controls.
- ❏ Verify clean elements in humidistats when the system is started.
- ❏ Evaluate pneumatic systems for air tightness, restrictions caused by flattening of the tubing, and cleanliness of the system.
- ❏ Inspect electronic systems for grounded shielded cable, and location of amplifiers with respect to magnetic fields, such as large transformers.
- ❏ View graphic panels for damaged plastic, dirt between plastic and back plate, lacing of control wires and access for service to all controls.
- ❏ Verify control instructions, including sequence of operations, and control drawing furnished by the contractor when conducting final acceptance test. Check each function of the controls.

TESTING

Submittals

Be sure the contractor obtains approval of test procedures and other pertinent information prior to testing.

Procedures

- ❏ Make a record of all tests, including such information as who attended, methods and procedures of test, results, and conclusions. Check specifications to determine that contractor is recording sufficient data to comply with requirements.
- ❏ Before tests are scheduled, see that contractor has proper tools, equipment, and instruments, and gages should be certified and pretested.
- ❏ See that equipment is thoroughly checked and prepared for tests.
- ❏ Make sure strainers and filters are clean immediately prior to the test.

Types

- ❏ Check the testing of refrigerant piping. See that specified pressure is put on the lines. Make sure all joints are checked, and that leaks are detected, repaired, and retested until found satisfactory. Isolate all items which may be damaged by high pressure.

❏ See that hydrostatic test is performed on all water piping. Carefully check to see if there is a loss in pressure during the test.

❏ See that a performance test on the system is run for the duration specified. Make sure needed corrections and adjustments are made as determined during test.

❏ See that the contractor records all data required for the performance test.

❏ After successful tests, install a new oil charge in compressor. Change oil filters and socks, and provide a new cartridge in the refrigerant drier. (An oil charge is not required for factory sealed units.)

PAINTING

❏ See that the equipment is furnished with the correct finish. Watch for abrasions.

❏ Watch for miscellaneous ferrous metal items that are not primed.

❏ Require touching up, priming, and finish painting as specified.

OPERATING INSTRUCTIONS AND GUARANTIES

❏ See that equipment guaranties, schematic flow diagrams, and instructions for the operation of equipment are furnished and posted.

❏ Arrange for future operating personnel to be instructed on the operation of equipment. Make a record of instruction periods, including any complications, instructing personnel, and personnel instructed. Videotaping such training is a good practice.

TABLE 12-5 MECHANICAL LIFE SPANS	
Equipment Item	**Years**
Air-cooled condensers	20
Evaporative condensers	20
Insulation	
Molded	20
Blanket	24
Pumps	
Base-mounted	20
Pipe-mounted	10
Sump and well	10
Condensate	15
Reciprocating engines	20
Steam turbines	30
Electric motors	10
Motor starters	17
Electric transformers	30
Controls	
Pneumatic	20
Electric	16
Electronic	15
Valve actuators	
Hydraulic	15
Pneumatic	20
Self-contained	10
	(continued)

REFRIG./
AC

TABLE 12-5 (CONTINUED)	
Equipment Item	**Years**
Absorption	
Air conditioners	
Window unit	10
Residential single or split package	15
Commercial through-the-wall	15
Water-cooled package	15
Heat pumps	
Residential air-to-air	10
Commercial air-to-air	15
Commercial water-to-air	19
Roof-top air conditioners	
Single-zone	15
Multizone	15
Boilers, hot water (steam)	
Steel water-tube	24
Steel fire-tube	25
Cast iron	35
Electric	15
Burners	21
Furnaces	
Gas- oil-fired	18
Unit heaters	
Gas electric	13
Hot water or steam	20
Radiant heaters	
Electric	10
Hot water, steam	25

Equipment Item	Years
Air terminals	
Diffusers, grilles, and registers	27
Induction and fan-coil unit	20
V A V and double-duct boxes	20
Air washers	17
Ductwork	30
Dampers	20
Fans	
Centrifugal	25
Axial	20
Propeller	15
Ventilating roof-mounted	20
Coils	
Water, or steam	20
Electric	15
Heat exchangers	
Shell-and-tube	24
Reciprocating compressors	20
Package chillers	
Reciprocating	20
Centrifugal	23
Cooling towers	23
Galvanized metal	20
Wood	20
Ceramic	34

HVAC / ENERGY DEFINITIONS

Annual Fuel Utilization Efficiency Factor (AFUE)   The ratio of annual output energy to annual input energy which includes any nonheating pilot input loss for gas- or oil-fired furnaces or boilers. It does not include electrical energy.

British thermal unit (BTU) Btu   is the standard of measurement for heat energy. One Btu is the amount of heat to raise the temperature of one pound of water one degree Fahrenheit.

Cooling degree day (CDD)   Cooling degree days are based upon a difference in temperature and time. Usually it is an indicator in measuring the energy consumption in the region that you are currently building. It is basically the difference between the mean temperature of the region (usually 65°) and those days with a temperature different from that standard. For example, a day with a temperature of 76° has a mean temperature of 11 CDD. 76° minus the mean of 65° equals 11. Thus, the annual degree-days would be the total sum of days difference to reach a total for the entire year.

Coefficient of Performance (COP)—Cooling   The ratio of the rate of heat removal to the rate of energy input in consistent units, for a complete cooling system.

Coefficient of Performance (COP)—Heating The ratio of the rate of heat delivered to the rate of energy input in consistent units, for a complete heat pump system.

Coefficient of performance (COP) COP is the ratio between heat output and power input. For example, one kW of straight electric heat provides 3412 btuh. If a heat pump's COP is 3.00, the same kW delivers 10,236 Btuh. The COP ratio is calculated by dividing the total heat pump heating capacity by the total electrical input (W) and multiplying the result by 3.412.

$$Btuh/(W)\ 3.412 = COP$$

or $$Btuh/(kw)\ 3,412 = COP$$

Energy Efficiency Ratio (EER) The ratio of net equipment cooling capacity in Btu/h (W) to total rate of electric input in watts under designed operating conditions.

Heating Seasonal Performance Factor (HSPF) The total heating output of a heat pump during its normal annual usage period for heating, in Btu, divided by the total electric energy input for the same period, in watt hours.

REFRIG./
AC

Heating degree day (hdd)   As with the cooling degree day (CDD), the heating degree day is also used in estimating the amount of energy consumed as opposed to the amount of energy consumed for cooling. As with CDD, take the difference between the actual temperature and the mean temperature and add the sums of the days together to get the annual HDD days for the calendar year.

Seasonal Energy Efficiency Ratio (SEER)   The total cooling output of an air conditioner during its normal annual usage period for cooling, in Btu/h (W), divided by the total electric energy input during the same period of time in watt hours.

Seasonal energy efficiency ratio (SEER)   SEER is the total cooling of an air conditioner or heat pump in BTUs during its normal annual usage period.

THINK SAFETY AT ALL TIMES

MECHANICAL GLOSSARY

Accessible   Signifies access that requires the removal of an access panel or similar removable obstruction.

Accessible, readily   Signifies access without the necessity for removing a panel or similar obstruction.

Air circulation, forced   A means of providing space conditioning utilizing movement of air through ducts or plenums by mechanical means.

Air-conditioning system   An air-conditioning system consists of heat exchangers, blowers, filters, supply, exhaust and return-air systems and shall include any apparatus installed in connection therewith.

Alteration   A change in an air-conditioning, heating, ventilating or refrigeration system that involves an extension, addition or change to the arrangement, type or purpose of the original installation.

Appliance   A device or apparatus that is manufactured and designed to utilize energy and for which this code provides specific requirements.

Boiler, hot water heating A self-contained appliance from which hot water is circulated for heating purposes and then returned to the boiler, and which operates at water pressures not exceeding 160 lb/in^2 gage (psig) (1102 kPa gage) and at water temperatures not exceeding 250F (121°c.) near the boiler outlet.

Brazed joints A joint obtained by the joining of metal parts with metals or alloys that melt at a temperature above 1000 F (538°c.) but lower than the melting temperature of the parts to be joined.

Btulh The listed maximum capacity of any appliance, absorption unit or burner expressed in British thermal units input per hour.

Chimney One or more passageways, vertical or nearly so, for conveying flue gases to the outside atmosphere (See also "Vent").

Chimney connector A pipe that connects a fuel-burning appliance to a chimney.

Closet A small room or chamber used for storage.

Combustible material Any material not defined as noncombustible.

Combustion air The air provided to fuel-burning equipment including air for fuel combustion, draft hood dilution and ventilation of the equipment enclosure.

Concealed gas piping   Piping that is enclosed in the building construction without means of access.

Condensate   The liquid that separates from a gas due to a reduction in temperature, e.g., water that condenses from flue gases and water that condenses from air circulating through the cooling coil in air-conditioning equipment.

Condensing appliance   An appliance that condenses water generated by the burning of fuels.

Conditioned air   Air treated to control its temperature, relative humidity or quality.

Conditioned space   The space contained within an insulated building enclosure which is conditioned directly or indirectly by heating or cooling systems.

Confined space   A room or space having a volume less than 50 cubic feet per 1000 Btu/h of the aggregate input rating of all fuel-burning appliances installed in that space.

Control, limit   An automatic control responsive to changes in liquid flow or level, pressure, or temperature for limiting the operation of an appliance.

Control, primary safety A safety control responsive directly to flame properties that senses the presence or absence of flame and, in the event of ignition failure or unintentional flame extinguishment, automatically causes shutdown of mechanical equipment.

Convector A system incorporating a heating element in an enclosure in which air enters an opening below the heating element, is heated and leaves the enclosure through an opening located above the heating element.

Convenience outlet, gas A permanently mounted hand-operated device for connecting and disconnecting an appliance to the gas supply piping conforming to AGA 7. The device includes an integral, manually operated gas valve so that the appliance is capable of being disconnected only when the valve is in the closed position.

Damper, volume A device that will restrict, retard, or direct the flow of air in any duct, or the products of combustion of heat-producing equipment, vent connector, vent, or chimney.

Decorative gas appliance, vented A vented appliance installed for the aesthetic effect of the flames rather than functional effects.

Decorative gas appliances for installation in vented fireplaces   A vented gas-fired appliance designed for installation within the fire chamber of a vented fireplace, wherein the primary function lies in the aesthetic effect of the flames.

Dilution air   Air that enters a draft hood or draft regulator and mixes with flue gases.

Direct-vent appliance   A fuel-burning appliance with a sealed combustion system that draws all air for combustion from the outside atmosphere and discharges all flue gases to the outside atmosphere.

Draft   The flow of gases or air through chimney, flue, or equipment caused by pressure differences.

Mechanical or induced   The draft developed by fan, air, steam jet, or other mechanical means.

Natural   The draft developed by the difference in temperature of hot gases and outside atmosphere.

Draft hood   A device built into an appliance, or a part of the vent connector from an appliance, which is designed to (1) provide for the ready escape of the flue gases from the appliance in the event of no draft, backdraft or stoppage beyond the draft hood, (2) prevent a backdraft from entering the appliance, and (3) neutralize the effect of stack action of the chimney or gas vent on the operation of the appliance.

Draft regulator   A device that functions to maintain a desired draft in the appliance by automatically reducing the draft to the desired value.

Duct system   A duct system is a continuous passageway for the transmission of air which, in addition to ducts, includes but is not limited to duct fittings, dampers, plenums, fans and accessory air-handling equipment.

Equipment   A general term including materials, fittings, devices, appliances, and an apparatus used as part of or in connection with installations regulated by this code.

Evaporative cooler   A device used for reducing air temperature by the process of evaporating water into an airstream.

Excess air   Air that passes through the combustion chamber and the appliance flue in excess of that which is theoretically required for complete combustion.

Exhaust hood, full opening   An exhaust

hood with an opening at least equal to the diameter of the connecting vent.

Factory-built chimney A chimney composed of listed and labeled factory-built components assembled in accordance with the manufacturer's installation instructions to form the completed chimney.

Fireplace An assembly consisting of a hearth and fire chamber of noncombustible material and provided with a chimney, for use with solid fuels.

Factory-built fireplace A listed and labeled fireplace and chimney system composed of factory-made components, and assembled in the field in accordance with the manufacturer's instructions and the conditions of the listing.

Masonry fireplace A field-constructed fireplace composed of solid masonry units, bricks, stones or concrete.

Fireplace stove A freestanding, chimney-connected solid-fuel-burning heater with or without doors connected to the chimney.

Flame-spread index A numerical index indicating the relative surface-burning behavior of a material tested in accordance with ASTM E 84.

Floor furnace A self-contained furnace suspended from the floor of the space being heated, taking air for combustion from outside such space, and with means for lighting the appliance from such space.

Flue   See "Vent."

Flue, appliance   The passages within an appliance through which combustion products pass from the combustion chamber to the flue collar.

Flue collar   The portion of a fuel-burning appliance designed for the attachment of a draft hood, vent connector or venting system.

Flue gases   Products of combustion plus excess air in appliance flues or heat exchangers.

Fuel-piping system   All piping, tubing, valves and fittings used to connect fuel utilization equipment to the point of fuel delivery.

Furnace, warm-air   A vented heating appliance designed or arranged to discharge heated air into a conditioned space.

Gas   Fuel gas, such as natural gas, manufactured gas, undiluted liquefied petroleum gas (vapor phase only), liquefied petroleum gas-air mixture or mixtures of these gases.

Hazardous location   Any location considered to be a fire hazard for flammable vapors, dust, combustible fibers or other highly combustible substances.

Heat pump   An appliance having heating or heating/cooling capability and which uses refrigerants to extract heat from air, liquid or other sources.

High-temperature (H.T.) chimney   A high-temperature chimney complying with the requirements of UL 103. A Type H.T. chimney is identifiable by the markings "Type H.T." on each chimney pipe section.

Labeled   Devices, equipment, or materials to which have been affixed a label, seal, symbol, or other identifying mark of a testing laboratory, inspection agency, or other organization concerned with product evaluation that maintains periodic inspection of the production of the above labeled items which attests to compliance with a specific standard.

Listed and **Listing**   Terms referring to equipment which is shown in a list published by an approved testing agency qualified and equipped for experimental testing and maintaining an adequate periodic inspection of current productions and whose listing states that the equipment complies with nationally recognized standards when installed in accordance with the manufacturer's installation instructions.

Log lighter, gas-fired An unlisted manually operated gas-fired solid-fuel ignition device for installation in a vented solid-fuel-burning fireplace.

Low-pressure gas supply system A gas supply system with gas pressure at or below 0.5 psig (3.44-kPa gage).

LP gas Liquefied petroleum gas composed predominately of propane, propylene, butanes or butylenes, or mixtures thereof which are gaseous under normal atmospheric conditions, but are capable of being liquefied under moderate pressure at nonnal temperatures.

Manufacturer's installation instructions Printed instructions included with equipment as part of the conditions of listing and labeling.

Masonry chimney A field-constructed chimney of masonry units, bricks, stones, labeled masonry chimney units, or reinforced portland cement concrete, lined with suitable chimney flue liners.

Mechanical exhaust system Equipment installed in a venting system to provide an induced draft.

Mechanical system A system specifically addressed and regulated in this code and composed of components, devices, appliances and equipment.

Medium-pressure gas supply systems A gas supply system with gas pressure exceeding 0.5 psig (3.44 kPa gage) but not exceeding 5 psig (34 kPa gage).

Noncombustible material Materials that pass the test procedure for defining noncombustibility of elementary materials set forth in ASTM E 136.

Nonconditioned space A space that is isolated from conditioned space by insulated walls, floors, or ceilings.

Pellet fuel-burning appliance A closed combustion, vented appliance equipped with a fuel feed mechanism for burning processed pellets of solid fuel of a specified size and composition.

Pellet vent A listed vent conforming to the pellet vent requirements of UL 641 for venting pellet fuel-burning appliances listed for use with pellet vents.

Plenum A chamber which forms part of an air-circulation system other than the occupied space being conditioned.

Purge To clear of air, gas or other foreign substances.

Quick-disconnect device A hand-operated device that provides a means for connecting and disconnecting an appliance to a gas supply and that is equipped with an automatic means to shut off the gas supply when the device is disconnected.

Refrigerant A substance used to produce refrigeration by its expansion or evaporation.

Refrigerant compressor A specific machine, with or without accessories, for compressing a given refrigerant vapor.

Refrigerating system A combination of interconnected parts forming a closed circuit in which refrigerant is circulated for the purpose of extracting, then rejecting, heat. A direct refrigerating system is one in which the evaporator or condenser of the refrigerating system is in direct contact with the air or other substances to be cooled or heated. An indirect refrigerating system is one in which a secondary coolant cooled or heated by the refrigerating system is circulated to the air or other substance to be cooled or heated.

Regulator A device for reducing, controlling and maintaining the pressure in a portion of a piping system downstream of the device.

Regulator vent The opening in the atmospheric side of the regulator housing permitting the movement of air to compensate for the movement of the regulator diaphragm.

Return air Air removed from a conditioned space through openings, ducts, plenums or concealed spaces to the heat exchanger of a heating, cooling, or ventilating system.

Room heater A freestanding heating appliance installed in the space being heated and not connected to ducts.

Service piping The piping and equipment between the street gas main and the gas-piping system inlet, which is installed by and is under the control and maintenance of the serving gas supplier.

Smoke-developed index A numerical index indicating the relative density of smoke produced by burning assigned to a material tested in accordance with ASTM E 84.

Supply air Air delivered to a conditioned space through ducts or plenums from the heat exchanger of a heating, cooling, or ventilating system.

Type B vent A listed and labeled vent conforming to UL 441 for venting gas appliances with draft hoods and other gas appliances listed for use with Type B vents.

Type BW vent   A listed and labeled vent conforming to UL 441 for venting gas-fired vented wall furnaces listed for use with Type BW vents.

Type L vent   A listed and labeled vent conforming to UL 641 for venting oil-burning appliances listed for use with Type L vents or with listed gas appliances.

Unconfined space   A space having a volume not less than 50 ft^3 (1.42 m^3) per 1000 Btu/h (293 W) of the aggregate input rating of all appliances installed in that space. Rooms communicating directly with the space in which the appliances are installed, through openings not furnished with doors, are considered a part of the unconfined space.

Unusually tight construction   Construction in which:

1. Walls and ceilings exposed to the outside atmosphere have a continuous water vapor retarder with a rating of 1 perm [57 mg/(s. m^2 Pa)] or less with openings gasketed or sealed, and

2. Weather-stripping has been added on operable windows and doors, and Caulking or sealant are applied to areas such as joints around window and door frames between sole plates and floors, between wall-ceiling joints, between wall panels,

at penetrations for plumbing, electrical and gas lines, and at other openings.

Vent A passageway for conveying flue gases from fuel-fired appliances, or their vent connectors, to the outside atmosphere.

Vent connector That portion of a venting system that connects the flue collar or draft hood of an appliance to a vent.

Vent damper device, automatic A device intended for installation in the venting system, in the outlet of or downstream of the appliance draft hood, of an individual, automatically operated fuel-burning appliance and which is designed to automatically open the venting system when the appliance is in operation and to automatically close off the venting system when the appliance is in a standby or shutdown condition.

Vent gases Products of combustion from fuel-burning appliances, plus excess air and dilution air, in the venting system above the draft hood or draft regulator.

Vented gas appliance categories The following categories are used to differentiate gas utilization equipment according to vent pressure and flue gas temperature.

Category I An appliance that operates with a non-positive vent connector pressure and with a flue gas temperature at least 140°F (60°C) above its dewpoint.

Category II   An appliance that operates with a nonpositive vent connector pressure and with a flue gas temperature less than 140°F (60°C) above its dewpoint.

Category III   An appliance that operates with a positive vent pressure and with a flue gas temperature at least 140°F (60°C) above its dewpoint.

Category IV   An appliance that operates with a positive vent pressure and with a flue gas temperature less than 140°F (60°C) above its dewpoint.

Ventilation   The process of supplying or removing conditioned or unconditioned air by natural or mechanical means to or from any space.

Venting   Removal of combustion products to the outdoors.

Water heater   A closed vessel in which water is heated by the combustion of fuels, electricity or other energy source and withdrawn for use external to the vessel at pressures not exceeding 160 psig (1103 kPa gage), including the apparatus by which heat is generated and all controls and devices necessary to prevent water temperatures from exceeding 210°F. (99°C).

NOTES